T0253833

Undergraduate Lecture Notes in Physics

Undergraduate Lecture Notes in Physics (ULNP) publishes authoritative texts covering topics throughout pure and applied physics. Each title in the series is suitable as a basis for undergraduate instruction, typically containing practice problems, worked examples, chapter summaries, and suggestions for further reading.

ULNP titles must provide at least one of the following:

- An exceptionally clear and concise treatment of a standard undergraduate subject.
- A solid undergraduate-level introduction to a graduate, advanced, or non-standard subject.
- A novel perspective or an unusual approach to teaching a subject.

ULNP especially encourages new, original, and idiosyncratic approaches to physics teaching at the undergraduate level.

The purpose of ULNP is to provide intriguing, absorbing books that will continue to be the reader's preferred reference throughout their academic career.

Series editors

Neil Ashby
Professor Emeritus, University of Colorado, Boulder, CO, USA

William Brantley
Professor, Furman University, Greenville, SC, USA

Michael Fowler
Professor, University of Virginia, Charlottesville, VA, USA

Morten Hjorth-Jensen
Professor, University of Oslo, Oslo, Norway

Michael Inglis
Professor, SUNY Suffolk County Community College, Long Island, NY, USA

Heinz Klose
Professor Emeritus, Humboldt University Berlin, Germany

Helmy Sherif
Professor, University of Alberta, Edmonton, AB, Canada

More information about this series at http://www.springer.com/series/8917

Kurt Fischer

Relativity for Everyone

How Space-Time Bends

Second Edition

 Springer

Kurt Fischer
Department of Mechanical and Electrical
  Engineering
Tokuyama College of Technology
Shunan-Shi
Japan

ISSN 2192-4791   ISSN 2192-4805   (electronic)
Undergraduate Lecture Notes in Physics
ISBN 978-3-319-17890-5   ISBN 978-3-319-17891-2   (eBook)
DOI 10.1007/978-3-319-17891-2

Library of Congress Control Number: 2015943059

Springer Cham Heidelberg New York Dordrecht London

Printed on acid-free paper

Springer International Publishing AG Switzerland is part of Springer Science+Business Media
(www.springer.com)

To Yukiko
You edited the text to become readable

Preface

Dear Reader,

When I was a teenager, I first encountered the theory of relativity in the public library of my hometown. There were two kinds of books in this subject. The easy ones did not really explain but displayed large colored pictures showing some kind of science fiction. The serious books seemed to explain things that nevertheless remained hidden behind a mess of mathematical symbols and as a result they did not *really* explain. So I was back to square one. Nevertheless, it aroused my interest in physics. It also motivated me to fill the gap. The result is this book.

This book is about light, energy, mass, space, time, and gravity: it is through these concepts that we explain the theory of special relativity and the theory of gravity, known as the **theory of general relativity**.

We will use many **thought experiments** and show how physicists create and solve models. This method is the one used by Einstein himself. He understood the theory through both *physical* and *geometrical* pictures. We will follow Einstein's original train of thoughts as closely as possible, even using some of his own thought experiments.

We will present some involved arguments. Nevertheless you only need your imagination but no complicated mathematics to understand the essence of it all. However, the beauty of physics is that we *can* calculate the numbers. Therefore, we present the equations of the theory of relativity together with the most important *exact* solutions only by using elementary mathematics. Even the Einstein equation of gravity showing the bending of space and time, we present and solve *in detail* and in common language. By the end, we will see why the theory of general relativity is the *simplest theory* of gravity.

That is to say there is a common misconception about general relativity: that it is impossible to understand without higher mathematics, and that it is therefore only for a few experts. However, already in 1973 the famous textbook "Gravitation"[1] advises on how to explain general relativity.

> Only three basic principles are invoked: special relativity physics, the equivalence principle, and the local nature of physics. They are simple and clear. To apply them, however, imposes a double task: (A) take space-time apart into locally flat pieces (where the principles are valid), and (B) put these pieces together again into a comprehensible picture. To undertake this dissection and reconstruction, to see curved dynamic space-time inescapably take form, and to see the consequences for physics: that is general relativity.

I am convinced, and believe that you will be too, that the book you hold in your hands realizes this concept.

Synopsis of the Contents

In the first four chapters, we explain what is called the theory of special relativity. We describe the relation between light, matter, space, and time.

1. In Chap. 1, we introduce the basics and describe that mass and energy are the opposite side of the same coin.
2. In Chap. 2, we will see why time and length are "relative."
3. In Chap. 3, we will see why any current-carrying wire exhibits relativity in everyday life.
4. In Chap. 4, we learn that, while riding a merry-go-round, school geometry is no longer true.

In Chaps. 5–9, we explain the theory of general relativity which describes gravity.

5. In Chap. 5, we show that Earth's gravity does not pull at all, but rather that it bends space *and* time.
6. In Chap. 6, we will see in detail the effects of the bending of space and time.
7. In Chap. 7, we thoroughly explain the meaning of the famous Einstein equation of gravity and the reason why it is the simplest possible way of describing gravity.
8. In Chap. 8, we introduce the most famous *exact* solution of the Einstein equation, i.e., the Schwarzschild solution, in simple terms.
9. In Chap. 9, we use the solutions of the Einstein equation and explore the famous predictions of the theory of general relativity, e.g., how much a light beam bends while passing at the sun, how large and heavy black holes are, why the orbits of the planets revolve slowly around the sun, and why the universe had a Big Bang but why its future is unclear to us.

---

[1]C.W. Misner, K.S. Thorne, and J.A. Wheeler: *Gravitation*, Freeman (1973)

Units and Symbols

One remark on highlighted words: indexed words appear in **boldface.** This means you can more easily find them on the corresponding pages.

Before describing the strange properties of light, we first explain units for measuring and compare the size of things. In physics, we use certain **units** to measure things. We measure lengths only in meters, time only in seconds, *not* in minutes, hours, or days. We measure mass only in kilograms. Any *other* unit used in this book is a combination of these units. For example, speed is measured in meters per second. Other units such as pounds or inches or the like are never used. The merit of this is that we can drop the units in all calculations because we know anyway which units are to be added *afterward* and we *fixed* them at the beginning.

We will often encounter really large or very small numbers. For example, while numbers like "one thousand" can be written down as 1000, the number one billion two hundred fifty-two million seven hundred eighty-four (1,250,000,784) is much harder to read. Usually, we are only interested in the first three digits or so for a rough estimate of how large things are. Here physicists count the number of digits after the first one—that is nine in this case—and they write

$$1.25 \times 10^9$$

In the same way, we write a very small number like 0.000145 as 1.45×10^{-4} by counting the number of leading zeros. Then we can easily multiply such numbers: we multiply $1.25 \times 10^9 \times 1.45 \times 10^{-4}$ by first multiplying 1.25 and 1.45 which is roughly 1.81, and adding the exponents $9 + (-4) = 5$. The result is $\approx 1.81 \times 10^5$. Here the symbol $\approx$ means "**is roughly equal to**".

For example, for the speed of light we usually use the rough value

$$\text{speed of light} \approx 3.00 \times 10^8 \text{ meters per second} \qquad (1)$$

We collect other important numbers appearing throughout the text in Table A.1 for reference.

Special Wording

And lastly, one remark about how we use phrases like "far enough away from something," "fast enough," and the like. For example, we say

> if the astronaut is far enough away from Earth, the astronaut can nearly neglect Earth's gravity.

We are aware of the fact that gravity is *never* exactly zero, even *very* far away from Earth. The fact of the matter is that the astronaut wants to measure some effect without gravity spoiling the measurement by the one percent of the result that the astronaut would get in a *really* empty space. So, if the astronaut thinks that gravity is still spoiling his experiment too much, he is *free to move* to a place which is *so far away* from Earth that there gravity *does* really spoil his experiment only up to the desired one percent at the most. Of course, if he wants to measure even more accurately, he must move even farther away from Earth. In short, if he moves "far enough away" from Earth, he can always neglect Earth's gravity *to the extent that he wants to* neglect it.

Tokuyama, Japan   Kurt Fischer
April 2015

Contents

1 Light, Matter, and Energy   1
   1.1   Light Beams.   1
   1.2   First Law of Relativity: Straight, Steady Speed
           Is Relative   1
   1.3   Measuring the Speed of Light.   3
   1.4   Second Law of Relativity: The Speed of Light
           Is Absolute   4
   1.5   Faster than Light?.   5
   1.6   Theory and Practice   6
   1.7   Mass and Inertia.   6
   1.8   Inertia and Weight: A First Glance   7
   1.9   Energy   7
   1.10   Mass and Motion Energy.   8
   1.11   Resting-Mass and Motion Energy   9
           1.11.1   Internal Motion-Energy   9
           1.11.2   Pure Energy   9
   1.12   Inertia of Pure Energy   10
   1.13   Mass Is Energy Is Mass.   12
   1.14   Information Needs Energy for Transport   13

2 Light, Time, Mass, and Length   15
   2.1   Light and Time.   15
   2.2   The Gamma Factor   17
   2.3   Whose Clock Is Running More Slowly?   19
   2.4   Light, Time, and Length   20
           2.4.1   Length in the Direction of Movement   20
           2.4.2   Length at Right Angles to Movement   21
   2.5   At the Same Time?.   22
   2.6   Are There Any Time Machines?.   24
   2.7   Time and Mass.   25
   2.8   Speed Addition.   27

3   Light, Electricity, and Magnetism   29
   3.1   Electric Charge and Speed   29
   3.2   Electric Charges and Magnets   30
   3.3   Electric and Magnetic Fields   32
   3.4   Magnetic Field from Electric Current   33
            3.4.1   The Faraday Paradox   34
            3.4.2   No Attraction Without Relativity   35
            3.4.3   Attraction with Relativity   35

4   Acceleration and Inertia   37
   4.1   Rotating Motion: Twin Paradox 1   37
   4.2   Rotating Motion: Not School Geometry   39
   4.3   Straight Motion and Acceleration   41
   4.4   Proper Time and Inertia: Twin Paradox 2   42
   4.5   Inertial State, Acceleration, and Proper Time   44

5   Inertia and Gravity   45
   5.1   Gravity Is Not a Force   46
   5.2   Gravity Bends Space-Time   48
            5.2.1   Bended Surface   49
            5.2.2   Bended Space-Time   51
   5.3   Measuring the Bending of Space-Time   52

6   Equivalence Principle in Action   57
   6.1   Time and Gravity   57
   6.2   Proper Time in Bended Space-Time: Twin Paradox 3 ..   59
   6.3   Moving Straightly in Bended Space-Time   61
   6.4   Length Under the Gravity of a Perfect Ball   62
   6.5   Gravity Around a Perfect Ball   64
   6.6   Mass Under Gravity   65
   6.7   Light Under Gravity   66
   6.8   Black Holes: A First Look   68
   6.9   Equivalence Principle: Summary   69

7   How Mass Creates Gravity   71
   7.1   Gravity in a Lonely Cloud   71
   7.2   Einstein Equation of Gravity   73
   7.3   Enter Pressure   73
   7.4   Enter Speed   75
   7.5   Enter Outside Masses   76
   7.6   Local and Global Space-Time   76
   7.7   Bended Space-Time and Tensors   77
   7.8   How to Solve the Einstein Equation of Gravity   78

8 Solving the Einstein Equation of Gravity 81
   8.1   Gravity Causes the Law of Motion 81
   8.2   Gravity Inside a Perfect Ball of Mass 82
   8.3   Flat Space-Time Inside a Ball-Shaped Hollow 85
   8.4   Gravity Outside a Perfect Ball of Mass 86
   8.5   Schwarzschild Exact Solution....................... 88
   8.6   Newton Law of Gravity............................ 91

9 General Relativity in Action 93
   9.1   Black Holes..................................... 93
   9.2   Light Bending: Weak Gravity 1 95
   9.3   Kepler Laws.................................... 101
   9.4   Planet Orbits Rotate: Weak Gravity 2 104
   9.5   Strong Gravity Near Black Holes 108
   9.6   Gravitational Waves 111
   9.7   Where Is the Gravity Energy? 113
   9.8   Big Bang of the Universe 115
         9.8.1   Small Ball of Mass in the Universe............ 116
         9.8.2   Large Ball of Mass in the Universe............ 118
   9.9   Vacuum Energy and Gravity 121

10 Epilogue... 125

Appendix 127

Index 135

Chapter 1
Light, Matter, and Energy

1.1 Light Beams

An astronaut floating in space switches on a torch or a laser beamer. The emerging light beam travels at the speed

$$299,792,458 \text{ meters per second} \tag{1.1}$$

This is the **speed of light in vacuum**. *Exactly.*

How can we verify this? First of all, we need an apparatus to send a light beam such as the box on the left in Fig. 1.1. It is open to the right. This box symbolizes the torch, or the laser, or the *sender* for short. The black horizontal arrow stands for the light beam. It travels through the gray box on the right which is some speedometer *measuring* the speed of the light beam in meters per second. We will *not* discuss what constitutes such an apparatus: we just assume that there *are* such devices.

1.2 First Law of Relativity: Straight, Steady Speed Is Relative

Does the speed of light change if we move the torch while sending a light beam? This prompts the question: *is moving relative* to what?

If we are moving in a fast train, we do *not* feel the steady speed of the train but only feel a slight tremble: that is the *non-steady* part of the speed. For example, in Fig. 1.2 a table stands inside a train. The train is moving straightly and steadily. The black ball will not begin to move on the table. Do you feel the tremendous speed while sitting at a table at home and feeling to be at rest? Which speed? For example, we could be referring to the tremendous speed at which the Earth is moving around the sun at least during a few minutes, say nearly straight and steady, or the speed at

© Springer International Publishing Switzerland 2015
K. Fischer, *Relativity for Everyone*, Undergraduate Lecture Notes
in Physics, DOI 10.1007/978-3-319-17891-2_1

Fig. 1.1 Light beam from *left to right*, sketched as *black horizontal arrow*. The *box* on the *left* stands for some sender like a torch or a laser; the *gray box* on the *right* is the light-speedometer, displaying the speed of light

Fig. 1.2 The ball on the glass table in a train will not begin to move on the table, if the train moves straightly and steadily

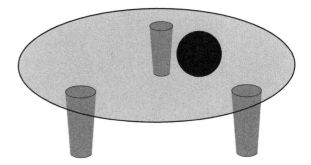

which the whole solar system is moving along the galaxy and/or the speed at which the galaxy is moving—to where?

Since Galilei at the end of the Middle Ages, we know that we *cannot* detect the speed if we are steadily moving straight ahead in some direction, by *any means whatsoever*. We can always take the point of view that *we are at rest* in the same way as we do when we "sit at a table". Even while sitting at table in a train that is steadily moving straight ahead, we can say that *we* are at rest and the whole station, plus the surrounding area we pass through, is moving *towards* us. At the same time, a friend standing on the platform of the station objects: of course the train is moving towards the station and the station together with the Earth around it are at rest!

What is correct? Answer: *both* we and our friend are right about insisting to be at rest: the train is moving only *relative* to the train station and the Earth around it. This is the **first law of the theory of relativity** formulated by **Galilei** a few hundred years ago:

> The speed of straight, steady motion of a body can only be measured *relative* to other bodies. The laws of nature do *not* depend on a straight and steady speed at which we may move relative to other bodies.

1.3 Measuring the Speed of Light

We measure the speed of the bodies in Fig. 1.3 *relative* to the ground. While sending the light beam, we move the torch towards the speedometer relative to the ground at say 10,000 meters per second, whereas the speedometer rests relative to the ground. To avoid that anything might disturb the light beam, we assume a vacuum, i.e. no air above the ground.

Nevertheless, the speedometer shows the very same speed of light! Now, this does not sound strange. Here is an analogy: replace the torch by a loudspeaker and the speedometer for light by a speedometer for sound, as in Fig. 1.4. We drew the sound as a white arrow. We depicted the calm air as light gray background. The sound travels through the calm air at a speed of about 343 meters per second.

The sound travels through the calm air and so its movement does not feel the moving loudspeaker. We will measure the same 343 meters per second, even if the loudspeaker travels towards the speedometer at 40 meters per second. So, is this not a very similar situation as the light beam?

No, next we will set the loudspeaker at rest inside the calm air and move the *speedometer* towards the loudspeaker at the same 40 meters per second.

Since the sound moves relative to the calm air to the right at 343 meters per second and the detector also moves relative to the calm air to the left at 40 meters per second, we will measure a sound speed of

$$343 + 40 = 383 \text{ meters per second.}$$

This is sketched in Fig. 1.5.

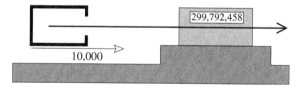

Fig. 1.3 Now the sender moves towards the speedometer at 10,000 meters per second relative to the ground. The ground is indicated in *dark gray*

Fig. 1.4 A loudspeaker moves towards the speedometer at 40 meters per second. The speed of sound remains the same as if the loudspeaker would rest

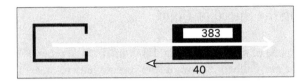

Fig. 1.5 If the speedometer is moving towards the loudspeaker which is resting in the air we measure a larger speed of sound

1.4 Second Law of Relativity: The Speed of Light Is Absolute

Next we repeat that thought experiment with light, as in Fig. 1.6.

The astonishing result is that for light the speedometer *still* shows the same speed of light! That means that light does *not* need any medium like "air" to travel. Light travels *as such* through the empty space. What is more, these thought experiments show that the speed of light is **absolute**. Therefore, as one of the *c*onstants of nature, it got its own name "**c**". Therefore, to simplify calculations, physicists recently adjusted finely the length of the meter such that the speed of light has exactly the value of Eq. (1.1).

> Light always travels at the same speed $c = 299,792,458$ meters per second through the vacuum. That is: **the speed of light is absolute**.

This phenomenon of nature is the **second law of the theory of relativity**. Physicists have tested and confirmed it in many experiments with ever growing precision over the last hundred years or so. It is the starting point for the **theory of relativity**.

In the next section we think about some consequences qualitatively, and then we introduce the more detailed physical concepts later on.

Fig. 1.6 If the speedometer moves towards the sender, the speed of light does not change

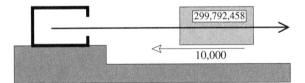

1.5 Faster than Light?

Clearly, there are things moving faster than sound. For example: when we hear the cracking sound of a whip in the circus, the tip of the whip is moving faster than sound! Another example: when using a fast airplane we can overtake the sound so that in the setup of Fig. 1.7 the sound does not reach the speedometer any more.

However, since light always travels at the same speed, we cannot overtake it. Suppose that in Fig. 1.8 we travel at $299, 792, 458 - 1$ meters per second to the right away from the torch.

Nevertheless we will see the light in the speedometer moving at the same speed as before. This means that we cannot escape from the light beam:

We *cannot* travel **faster than light!**

You think this is too weird and cannot be true? Here is another example. In everyday life we see that whatever is moving will come to rest if we do not sustain its movement: balls, cars, airplanes, and so on. However, in reality, bodies will move *forever* along a straight line at a steady speed, if nothing hampers them. What hampers them on Earth is that by friction they heat up other bodies and *as such* lose their energy and eventually come to rest. From an everyday point of view, this appears to be weird, but since a few hundred years, we know this is correct. So we should not judge from our everyday experience because that may be not satisfactory.

Fig. 1.7 Escaping from sound

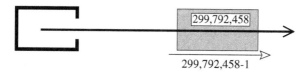

Fig. 1.8 Even if we try to escape from a light beam at nearly the speed of light, it eventually passes us at the speed of light c

Fig. 1.9 Trying to
accelerate a peppercorn by
putting it on *top* of a rocket

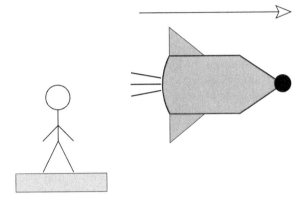

1.6 Theory and Practice

Any engineer should take the statement that we cannot travel faster than light as a
challenge. Let us build a powerful rocket which converts nearly all its fuel to thrust.
The payload shall be just a black peppercorn. Now start the rocket and observe the
peppercorn on top of the rocket accelerating, moving faster and faster away from us.
In Fig. 1.9, we sketched the peppercorn on top of the rocket larger than life-size to
get a better view.

All the same, the faster the peppercorn becomes, the more difficult for us to
accelerate it further. If it has reached 99 percent of the speed of light—that is, if
it is moving away from us at this speed—then it seems to hardly react at all to the
additional thrust of the rocket.

1.7 Mass and Inertia

This is the time to ask *what* has changed within the peppercorn that it resists further
acceleration so much? Its **inertial mass** has grown! You can test "inertial mass" that
is sometimes called for short **mass** or **inertia** by putting a polished stone on a skating
rink as sketched in Fig. 1.10. Even if the friction from contact with the ice is small, it
resists being "pushed", that is being accelerated. If the stone is twice as big, you have
to use twice the effort to move it. That's because the bigger the stone, the more mass
it contains. Clearly inertia has the connotation of "laziness": matter resists being
pushed because all kinds of matter have mass to a different extent. Experience tells
us that having mass is *the* fundamental property of matter.

Fig. 1.10 Pushing inertial mass on an ice skating rink

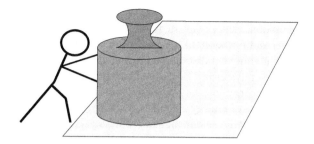

1.8 Inertia and Weight: A First Glance

Usually we measure mass of a say stone by *weighing* it. However, what on Earth should *heaviness* have to do with *massiveness*?

The massier a stone is, the *less* it tends to move. The *heavier* a stone is, the *more* it tends to move, that is to say, to fall down!

What is more: on Earth a stone has some weight but out in space—and most matter is out in space—we *cannot* weigh our stone because there is no planet conveniently at hand! However we can push it, try to **accelerate** it, and measure how much it resists.

Nevertheless there *is* a deep connection between mass and weight which we will explore later on in Chap. 5. For the time being, we continue to use the inertial mass.

Now to come back to our peppercorn: the faster it passes us, the more it resists being accelerated to a higher speed. This means that its *mass* must have increased. In general:

The faster a material body passes us, the more *mass* it contains.

However, nothing comes from nothing, so what did we add to the peppercorn that it became massier?

1.9 Energy

We burned fuel in the rocket and provided the peppercorn with what is called **motion energy** or **kinetic energy**, that is energy of motion. There are many kinds of energy: heat energy, electrical energy, or motion energy, but it turns out by careful experiments that no matter which type of energy it is, we can *convert* it into any other type. For example: while driving a car we convert the electrical energy residing in the chemical bonds of the gasoline molecules into motion energy of the molecules by burning the gasoline, and this into the motion energy of the car, and that is what we want. However, we carry away air which breaks away as air whirls behind the car and that part of our motion energy converts into motion energy of air whirls. These whirls break up further and further until the energy converts into heat through air friction. However, heat is nothing but disordered movement of molecules...

It took physicists a long time to establish by experiments that we can convert *any* form of energy into *any* other *without loss*. In fact, it took a few centuries before the **concept of energy** established itself: why is it that heat, electricity, or a fast car should basically contain the same stuff, that is to say *energy*? Nevertheless, physicists found in every experiment so far that energy may be converted from one into another but it never disappears or springs up from nothing.

We define one form of energy, say motion energy to specify the unit of energy, the **Joule**: two kilograms of mass traveling at the speed of 1 meter per second have about 1 Joule of motion energy.

Let us convert 1 Joule of energy into different forms in both directions. In the end, there is still just this 1 Joule of energy, just enough to accelerate the kilogram as described before. We stress this is an experimental fact which is not something that you can prove but only suggest from experience. Physicists say that

> **Energy is conserved**: we can convert energy into different forms but not create or destroy it.

To come back to our rocket: clearly the rocket has converted the main part of the energy being originally in its fuel into the motion energy of the peppercorn. If we burn out our rocket, we have supplied the peppercorn with some energy.

1.10 Mass and Motion Energy

How can we measure the energy of the peppercorn? Simply put a wall into its path and assess the damage: the larger the hole, the more energy the peppercorn brought along. Twice the powerful rocket will load the peppercorn with roughly twice as much energy. However, if the original rocket accelerated the peppercorn to 99 percent of the speed of light, then twice the strong rocket will achieve the peppercorn to fly at about 99.75 percent of c which is not much faster.

So, where is the energy going when the peppercorn cannot become much faster? Answer: the more matter approaches the speed of light, the more its motion energy shows up as mass.

In other words: the more motion energy matter carries, the more mass it has. That suggests that mass itself is a form of energy: can we imagine a stone to be a kind of frozen-in motion energy?

1.11 Resting-Mass and Motion Energy

We can convert the different types of energy into each other. What then about mass of a resting body, the **resting-mass**? Up to know, we imagined resting-mass as rigid as a stone. However, we know that mass consists of electrons, protons and the like which all the time move around each other. These small constituents are not at rest, even when the "stone" rests.

1.11.1 Internal Motion-Energy

We know that heat means the disordered motion of atoms, that is the motion energy of atoms. If we heat a stone it should contain more motion energy and hence more mass: the hotter, the massier! Here is a graphic example of internal motion. Take a box inside of which two identical balls bounce back and forth, as in Fig. 1.11.

Inside the box they will bounce back at exactly the same time to the left and the right wall so that the box remains resting on the ground. They carry motion energy and hence they carry *more* than just their resting-mass. The faster they bounce back and forth, the more *resting-mass* the box gains!

We will calculate how much more mass such a moving body has in Sect. 2.7.

Coming back to our resting stone, how much of the resting stone is actually motion energy of its atoms? We see that, by and by, the concept of "mass" and "energy" blend in. They seem to be two sides of the same coin!

1.11.2 Pure Energy

Motion energy is attached to a mass that can be at rest. This prompts the question: is there **pure energy** without **resting-mass**? If there is, it never can rest beside us. What resting energy would be left over without mass? However, we already learned of such thing. In fact, *light* is pure energy. It never rests because it moves always at the speed c, *no matter how much we change our speed*. This experimental fact was the starting point of the theory of relativity, and we see that the more we probe it, think about it, the more astonishing details it reveals. To coin a catch-phrase:

Fig. 1.11 Two internal bouncing masses increase the **resting-mass** of the box

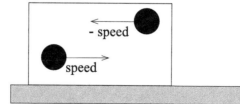

> **Light is pure energy**, without resting-mass, *rest-less*, always moving at speed c.

If mass and pure energy are really two sides of the same coin, we would then like to have a thought experiment connecting the two. In fact, Einstein himself provided one such thought experiment, the essence of which we present now.

1.12 Inertia of Pure Energy

Think of a wall standing on an ice skating rink, as in Fig. 1.12. The wall contains some light bulbs so that light emits from the wall to the right. It builds up a package of light which we sketched in light gray.

When the package has reached some width, we switch off the light bulbs so that it starts to leave the wall. After some time, the light package is at some distance from the wall, as we see in Fig. 1.13.

While energy is being built up inside the light-gray volume, it presses against the wall because that **pressure** *is* energy per volume. How can we understand that? Think of a pressure cooker. If it is under some pressure, it certainly contains energy. The energy is released by opening the lid of the cooker. Twice as much pressure means twice as much energy in the cooker, and the same pressure in a twice as large cooker means twice as much energy as well. Hence energy is pressure times volume because it doubles if we double the pressure or the volume. In other words: pressure is energy per volume.

Now we know that the package of light as energy presses against the wall before leaving it. In other words: the wall *recoils* from the leaving light package and moves to the left at some speed. However, this is weird. The massy wall suddenly begins to

Fig. 1.12 The packet of light has built up completely and starts to leave the wall. The *triangular bump* in the floor indicates the center of mass. The wall starts to recoil

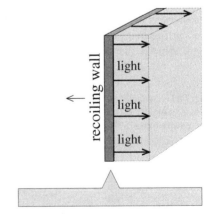

Fig. 1.13 The light packet has *left* the wall and moves at the speed of light to the *right*. The wall is slowly moving to the *left*

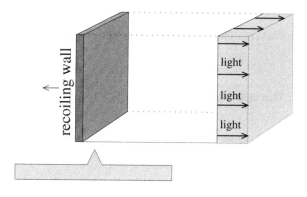

Fig. 1.14 The center of mass does not move without outside influence

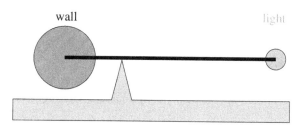

move to the left on the ice skating rink and nothing, that is no other mass is moving to the right? That is just not possible that the center of mass will not move without outside influence. There is only one way out: the light carries some mass to the right!

We sketched in Fig. 1.14 that if the large mass of the wall recoils a little to the left, then the light must carry a "light" mass over a longer distance to the right so that their masses are balanced again. This is similar to balanced weights on a scale, where their center of mass remains at the triangular bump.

How much mass does the light carry away? The wall was pushed by the light which is pure energy. If there goes twice as much light out to the right, then the wall recoils twice as much so that we expect that the light carries twice as much mass or inertia with it. We conclude that even a light package of **pure energy E** carries **mass m** and that the two are in proportion. The constant of proportion must be a constant of nature because we talk about a law of nature relating mass and energy. You find the detailed discussion in physical terms in Appendix A.2 showing that the constant of proportion is the inverse of the square of the speed of light:

$$m = \frac{E}{c^2} \tag{1.2}$$

In 1905 **Einstein** said[1]:

---

[1] A. Einstein. Ist die Trägheit eines Körpers von seinem Energieinhalt abhängig? Annalen der Physik, Volume 18, page 639, 1905. A. Einstein. Das Prinzip von der Erhaltung der Schwerpunktsbewegung und die Trägheit der Energie. Annalen der Physik, Volume 20, page 627, 1906.

Wenn die Theorie den Tatsachen entspricht, so überträgt die Strahlung Trägheit [...].

That is to say:

If the theory of relativity is correct, then radiation carries inertia.

By the way, in the original thought experiment Einstein made a mistake which we also explain in the Appendix A.2.

1.13 Mass Is Energy Is Mass

Let us sum up: the theory of special relativity *suggests* that

Inertial mass is a form of energy and **energy has mass**. Consequently there should be a way of converting resting-mass into energy and vice versa.

Physicists say "mass and energy are equivalent". We can rewrite the Eq. (1.2) as the famous formula

$$E = mc^2 \tag{1.3}$$

connecting energy, mass, and the speed of light.

We can see this formula (1.3) at work in Fig. 1.15 when light hits an atom. Then the atom recoils from the light, that is the energy of the light converts into motion energy of the atom. If the light carries enough energy, then sometimes some of its energy converts into an electron and a positron which are emitted from the much slower atom. So we can directly observe the conversion of energy into mass. By the way, the process works also in the opposite direction. We can then see how the matter of the electron and positron converts into pure energy.

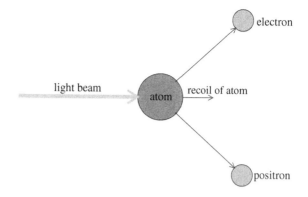

Fig. 1.15 Pure energy converts into matter in the form of an electron and a positron by hitting an atom

1.14 Information Needs Energy for Transport

We saw that material bodies like mass or pure energy cannot travel faster than light. However, what happens if we do not send any energy or mass but only the *information* about it? Maybe we can then send pure **information** faster than light by using some kind of telephone to send this information very far away to someone who puts together the mass or energy after our specifications.

However, it is just the other way round. Information needs mass or energy to travel! In order to handle information, you have to store it somehow to remember and therefore you need some mass or energy. Hence even pure information cannot travel faster than light.

Chapter 2
Light, Time, Mass, and Length

2.1 Light and Time

Let us think a little bit more about a light sender and a speedometer resting *relative* to each other, as in Fig. 2.1. Here we omit the speedometer. We put the whole setup into some transparent elevator as you find sometimes in department stores. We depicted it as a light-gray background box.

The man inside the elevator rests relative to the light-sender. He sees the light beam traveling horizontally. That's because the speed of the elevator has no absolute meaning to him, as we saw. He measures the time needed for the light beam to arrive at the tip of the arrow with the clock resting beside him.

Standing at the right outside the elevator, we will see the light starting at the sender. While the light moves to the right, we see that the elevator moves up together with the light beam. For example: at half the time it will be half the way up, so it just travels along the diagonal upwards.

This diagonal is *longer* than the horizontal distance. But remember that light travels always at the same speed c along the line whether measured inside or outside the elevator. We see that the **direction** of the light beam changes but the *magnitude* of the **speed of light** is for both of us the same. Hence we see the light traveling a *longer* time than the observer resting with the light source:

$$\begin{pmatrix} \text{time of clock} \\ \text{moving relative to us} \end{pmatrix} > \begin{pmatrix} \text{time of clock} \\ \text{resting relative to us} \end{pmatrix} \qquad (2.1)$$

That means that light and time are connected. For us outside, time inside the elevator evolves *more slowly* than ours. When the observer sitting inside the elevator says: "1 second has passed", we outside say: "No, *more* than 1 second has passed". In other words:

© Springer International Publishing Switzerland 2015
K. Fischer, *Relativity for Everyone*, Undergraduate Lecture Notes
in Physics, DOI 10.1007/978-3-319-17891-2_2

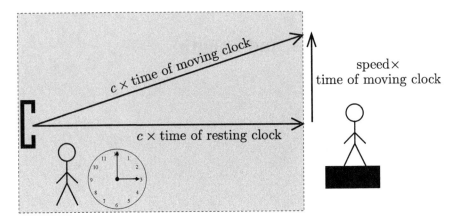

Fig. 2.1 We see the elevator moving upwards. Because speed is distance per time, we can write the distances as speed times time

> Because the magnitude of the **speed of light is absolute**, the pace of time is *relative* to the speed of the observer.

We use the Greek letter **gamma** "γ" to describe this: it denotes a number between zero and one and connects the larger time of the moving clock with the smaller time of the resting clock:

$$\left(\begin{array}{c} \text{time of clock} \\ \text{moving relative to us} \end{array} \right) \times \gamma = \left(\begin{array}{c} \text{time of clock} \\ \text{resting relative to us} \end{array} \right)$$

We see: for speed zero, γ is one because then we agree with the observer in the elevator on our times. The larger the speed of the elevator, the longer the diagonal becomes, so the smaller the factor γ.

Please have again a look at Fig. 2.1. During the time of the moving clock in which the light travels along the longer diagonal at a speed c, the elevator travels the *shorter vertical* line at its speed. Hence its speed is always *less* than the speed of light. We see again:

> Matter cannot travel **faster than the speed of light**, relative to us.

2.2 The Gamma Factor

We can actually compute the γ factor. We need only the **Pythagoras theorem** to get the answer. Please have a look at Fig. 2.2. We use the same triangle from Fig. 2.1. The theorem of Pythagoras tells us that

$$(c \times \text{time of moving clock})^2$$
$$= (c \times \text{time of resting clock})^2 + (\text{speed} \times \text{time of moving clock})^2$$

We want the time of the resting clock, looking from the outside, so

$$c^2 \times (\text{time of moving clock})^2 - \text{speed}^2 \times (\text{time of moving clock})^2$$
$$= c^2 \times (\text{time of resting clock})^2$$

Factor out the "time of the moving clock" on the left,

$$(\text{time of moving clock})^2 \times \left(c^2 - \text{speed}^2\right) = c^2 \times (\text{time of resting clock})^2$$

Then divide by c^2:

$$(\text{time of moving clock})^2 \left(1 - \frac{\text{speed}^2}{c^2}\right) = (\text{time of resting clock})^2$$

Because the square of the moving time and the resting time are never negative, the factor $\left(1 - \frac{\text{speed}^2}{c^2}\right)$ must be at least zero as well. Hence we see here again that the speed cannot be larger than the speed of light. Further, we can extract the square root to get the relation between our moving time and the resting time inside the elevator:

$$(\text{time of moving clock}) \times \sqrt{1 - \frac{\text{speed}^2}{c^2}} = \text{time of resting clock} \qquad (2.2)$$

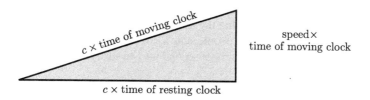

Fig. 2.2 The Pythagoras theorem tells us how the times of the resting and the moving clock depend on each other

The square root term is the γ **factor,**

$$\gamma = \sqrt{1 - \left(\frac{\text{speed}}{c}\right)^2} \tag{2.3}$$

We can also get the γ factor in a graphic way. We use again the triangle of Fig. 2.2 and express the lengths of all three sites as multiples of the longest side, that is we divide all three sides by ($c \times$ speed of moving clock). Then the longest side has length one, by construction. The lower side becomes

$$\frac{\cancel{c} \times (\text{time of resting clock})}{\cancel{c} \times (\text{time of moving clock})} = \gamma$$

and the right side becomes

$$\frac{\text{speed} \times \cancel{(\text{time of moving clock})}}{c \times \cancel{(\text{time of moving clock})}} = \frac{\text{speed}}{c}$$

In other words, γ and $\dfrac{\text{speed}}{c}$ are the coordinates of a point on a circle of radius one, as you can see in Fig. 2.3. Hence we can see the size of the γ factor also from this figure. We see again that as the speed decreases to zero, γ increases to one.

This factor γ appears nearly everywhere in the theory of relativity, once you start to calculate things. Sometimes it is easier to use γ than to use the speed.

How much differs the γ factor from one for ordinary speeds on Earth? An airplane travels at roughly 1000 kilometers per hour, which is one million meters per 3600 seconds, or roughly 300 meters per second, that is 3×10^2. The speed of light is 3×10^8, so the airplane travels roughly at 10^{-6}, that is one part of a million of the speed of light

$$\frac{\text{speed}}{c} \approx 10^{-6}$$

Now square this. It becomes practically zero, namely one part in a million-million, that is 10^{-12}. Hence the gamma factor is practically one for ordinary speeds on Earth. Can we conclude from this fact that the theory of relativity does not play any role for nature on Earth, as sometimes people state? No! We will see in Chap. 3 that even at speeds of less than one millimeter per second we can easily observe the time slip!

Fig. 2.3 The Pythagoras theorem tells us how the γ factor depends on the speed

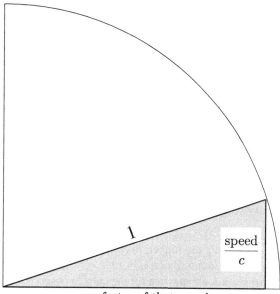

γ factor of that speed

For small speeds, we can estimate the γ factor, as you will find in Appendix A.3. The result is

$$\gamma = \sqrt{1 - \left(\frac{\text{speed}}{c}\right)^2} \approx 1 - \frac{1}{2}\left(\frac{\text{speed}}{c}\right)^2 \tag{2.4}$$

We list the most important **properties of the gamma factor**:

1. For zero speed, γ is one.
2. The larger the speed, the smaller γ.
3. For nearly the speed of light, γ is nearly zero.
4. For very small speed, γ is smaller than one by a factor which is in proportion to the square of the speed.

2.3 Whose Clock Is Running More Slowly?

Let us now place the light-sender *outside on the left of* the elevator and let us be inside, as in Fig. 2.4. Let the elevator move upwards. Then the outside observer will see the light beam passing horizontally and we see it passing downwards. As for us,

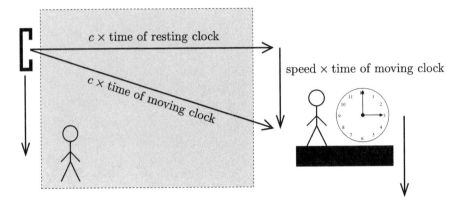

Fig. 2.4 We inside the elevator see the observer outside moving down, together with the light-sender

the light travels the longer, diagonal path and we conclude that for us the *time of the observer standing outside of the elevator* is running more slowly. However, in the last section we saw that for the outside observer the time *inside* the elevator runs more slowly than his own!

Compare this with Fig. 2.1. Then whose time is *really* running more slowly?

Answer: the question is wrong! It is the kind of question like "When is the air feeling colder: at night or outside?" We cannot decide because we cannot *compare* the two situations. The same goes for this situation: if the inside and outside observers want to compare their clocks, one of them or both have to *change* their speed to stop near each other. *Then* they can compare their clocks. However, by changing speed our clocks will change their pace! We will see in Chap. 4 what will happen then. For the time being, the observers pass each other and *continue* to travel at a steady speed. Hence *both* are correct in their statements that the other one's clock is going more slowly.

2.4 Light, Time, and Length

2.4.1 Length in the Direction of Movement

How do the observers measure the speed of the setup in Fig. 2.1? At first, we as the outside observer put down some rod in front of us. We sketch the rod as the solid black arrow pointing upwards in Fig. 2.5. Then we measure the length of the rod.

Because the rod rests relative to us, we call this the "length of resting rod". Then we measure the time needed for the setup to pass the rod. We choose the time of the *moving* clock. We get the speed

$$\text{speed} = \frac{\text{length of } resting \text{ rod}}{\text{time of } moving \text{ clock}}$$

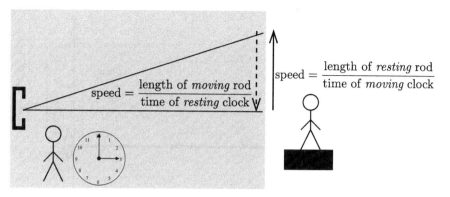

Fig. 2.5 Measuring the relative speed in terms of own length and time differences

What does the observer inside the setup measure? Because constant speed is relative, the observer can insist that *he* rests while the rod outside and we travel with at the same speed *downwards*. This is indicated by the dashed arrow pointing downwards. As a result, the observer measures the same speed in terms of the moving rod and clock,

$$\text{speed} = \frac{\text{length of } resting \text{ rod}}{\text{time of } moving \text{ clock}}$$

While for us outside the moving clock measures one second, the observer inside the elevator sees this clock resting and therefore showing only less than one second, namely $\gamma < 1$ seconds. Hence for him, the length of the *moving* rod must be *shorter* than for us, by the same γ factor, to get the same speed:

$$\left(\begin{array}{c}\textbf{length of rod moving}\\\textbf{in direction of speed}\end{array}\right) = \gamma \times \text{(length of resting rod)} \qquad (2.5)$$

2.4.2 Length at Right Angles to Movement

What happens at right angles to the movement? In Fig. 2.6, we drew also the rails on which the elevator moves up and down. We sketched two wheels and one axis. Both observers measure the *same* length for the axis.

Why? For the outside observer, the rails always rest. Their distance is just the length of the *resting* axis. If the elevator moves up, then for the observer inside the axis has still the same length because speed is only relative, and he can insist that he rests. If the outside observer measures another length of the axis than the resting length, this would then mean for him that the axis would be longer or shorter than

Fig. 2.6 The length of a rod at right angles to the movement does not change

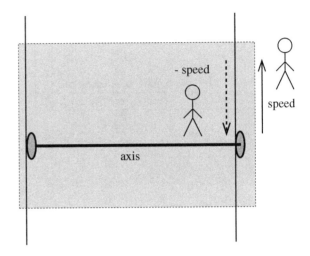

the distance of the still resting two rails: the elevator would derail! This is absurd. Hence:

Lengths at right angles to movement do not change.

2.5 At the Same Time?

Let us see how the relativity of time and length go hand in hand with the help of a famous thought experiment called "The pole and the barn". Here light does not enter, only a pole enters a barn and apparently contracts.

A friend has a gray pole slightly longer than the barn into which he wants to place it. The barn has an electrical front door at the left and an electrical back door at the right. At first, we check that the resting pole is really longer than the barn, as in Fig. 2.7.

Fig. 2.7 The resting pole is longer than the barn

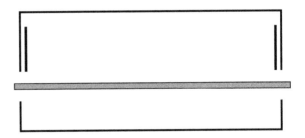

Fig. 2.8 The moving pole is shorter than the barn, for us standing inside the barn

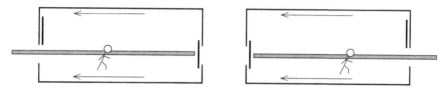

Fig. 2.9 For our running friend, the moving barn is shorter than the resting pole, but for him the right door closes and opens *before* the left door does

Then the friend takes the pole out of the barn and runs very fast through it from the left to the right. Because of the contraction described in Sect. 2.4.1 and for us standing in the barn, the pole looks *shorter* than the barn if the pole is only fast enough. We can arrange in advance that both doors close at the *same instant of time* and open shortly afterwards. If we choose the right moment for closing and opening the doors, the barn is *completely inside the barn* for a short moment as we see in Fig. 2.8.

However, what does our running friend see? The *barn* is moving towards him, so the *barn* is shorter for him. Hence the pole *does not fit* into the barn at all! By which miracle can the pole be completely *inside* the barn for some time, with both doors closed at least for a moment?

The point is that for the running friend the right door closes and opens again first and the left door closes and opens again afterwards, as you see in Fig. 2.9.

Why is that? Enter light. Suppose that some light flashes at the doors when pressing the switch to close both doors. As for us standing in the center of the barn, the flashes occur at the same time. Hence for us, the doors are closing *at the same time*.

What does our running friend see? He also sees the light of both flashes traveling at the speed of light c, as we know. However, the flash from the right door moves to the runner at a shorter distance because he runs *towards* it, and because he moves away from the flash coming from the left door. Hence he sees the right flash *before* the left flash: he sees that the right door is closing *before* the left door. The reason is twofold: first, the speed of light is absolute, and second, the two events of "the backdoor is closing" and "the front door is closing" happen at *different* places. We conclude:

The statement "**At the same time**" is not an absolute true statement for things happening at different places. Time depends on the relative speed of the clock and the observer.

2.6 Are There Any Time Machines?

Within the thought experiment of the pole and the barn, it seems to be possible to mix up the future and the past.

We can arrange to close the left door a little bit *before* the right door. If this "little bit" is short enough, we then have a strange situation which sounds as if a time machine will be possible.

For us standing in the center of the barn, the light from the left door reaches us *before* the light from the *right* door while our friend is running so fast towards the right door that *its* light reaches him *before* the light from the *left* door. In other words: as for us the left door closes *before* the right door while for our friend the left door closes *after* the right door.

Is it therefore possible to mix up the future and the past as in a time machine?

However, closing the left door does not *cause* the right door to close. Let us think about what will happen *if* the left door's closing *causes* the right door to close. Please have a look at Fig. 2.10.

When the left door begins to close, it sends light to us via the dashed arrow. At the same time, the left door sends a signal to the right door via the solid arrow. Via this signal it *causes* the right door to close, that is *when* the signal reaches the door. When the right door begins to close, this door itself sends some light to us via the dotted arrow.

We see the left door closing after the time it took light to travel *directly* via the dashed line to us. The signal causing the right door to close can travel no faster than the speed of light. Therefore we see the right door closing *at the earliest* after the time it took light to travel from the left door via the solid *and* dotted line to us. As

Fig. 2.10 If the left door's closing *causes* the right door to close we will see the left door's closing before the right door's closing, even if we are running in some direction relative to the doors

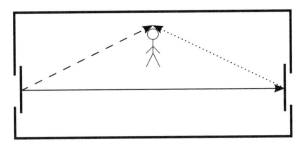

you see in the figure, the length of the dashed line is shorter than the length of the solid *plus* the length of the dotted line.

To sum up, we see the *cause before* the *effect*, that is the left door closing before the right door.

What does our running friend see? The runner sees the light traveling at the *same* speed of light, and also for him, the shortest path for the signal from the left door is the direct path. Hence he also sees that the left door closes before the right door!

In general: physical phenomena can not influence effects in the past. In other words: there is no such thing as a **time machine** "sending" us into the past, that is, influencing the past. Physicists call this the principle of **causality**.

2.7 Time and Mass

While we were standing outside the moving elevator in Sect. 2.1, we saw that time itself slows down for bodies inside that elevator. What are the consequences as to the bodies moving inside? It means that all movements slow down by exactly the same amount. So some property shared by all the bodies inside must change, at least from our standpoint. The obvious candidate is the inertia of the body. If they all become more inert, they move more sluggishly as if in slow motion. That fits with our observation before. For the same reason, when time slows down, the bodies will never be able to move faster than light moves from us away. To coin a catch-phrase:

> For bodies moving relative to us, time evolves more slowly than they experience themselves. Therefore they look more inert from our point of view. Their **inertia increases as their time** slows down, relative to us.

Let us see how much the inertia grows with speed. Suppose a ball with some resting-mass bounces at right angles, *slowly* against a wall and bounces back elastically as in Fig. 2.11.

When the ball bounces back, it changes from its original downwards speed to the same speed, but upwards. It makes no difference if the ball has say three times as much mass or if the same mass is moving three times as fast. So the "push" which the wall receives depends only on the product

$$push = mass \times speed$$

The wall receives *twice* this push: once when the wall absorbs the push from the ball and a second time when the wall pushes the ball elastically *back* at the same speed.

By the way, physicists call this push **momentum**.

Fig. 2.11 A ball bounces
against a resting wall

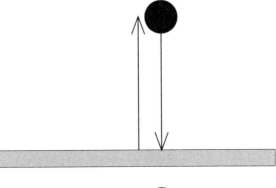

Fig. 2.12 The same ball
from Fig. 2.11 bouncing at
right angles from the same
wall, seen by us, moving
relative to the wall, along the
wall

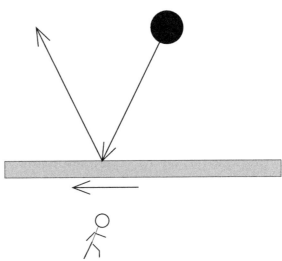

Further, the ball was slow enough so that its mass is by our experience nearly
unchanged, that is nearly the resting-mass. Hence the wall receives twice the
amount of

$$\text{push} = (\text{resting mass}) \times \text{speed}$$

Next, we imagine ourselves traveling fast *along* the wall, that is parallel to it, as in
Fig. 2.12. That is relative to us, the wall travels from the right to the left.

All the same, the wall will receive the same push at right angles. Also the distance
between the wall and the ball does not change for us because we learned in Sect. 2.4.2
that lengths at right angles to the motion do not change. The only difference is now
that we see the time of the ball proceeding more slowly by the factor γ. Hence we
see that the speed of the ball is *slower* by this factor. Therefore, to maintain the *same*
push, the mass of the ball must be *larger* by the same amount that its speed is *slower*,
that is by the same amount that its time is evolving more slowly:

$$\textbf{total mass} \text{ moving at speed} = \frac{\textbf{resting mass}}{\gamma} \tag{2.6}$$

We learned in Sect. 1.13 that this growing mass is due to the motion energy which the mass has gained. Therefore this **mass has the total energy**

$$\text{total energy of mass moving at some speed}$$
$$= \text{mass moving at speed} \times c^2 = \frac{\text{resting mass}}{\gamma} \times c^2 \tag{2.7}$$

2.8 Speed Addition

We already know that we cannot travel faster than light. In Sect. 1.6 we demonstrated that with a peppercorn driven by a rocket. However, maybe there is another method *without* needing to accelerate things? Here comes a thought experiment as shown in Fig. 2.13. In a very light, yet strong box there are two balls with equal resting-mass moving to the left and the right so that we can ignore the mass of the box and concentrate on the mass of the two balls.

This box moves at say 70 percent of the speed of light to the right, relative to the ground. Let the upper ball move at the speed of say 70 percent of the speed of light to the left in the box, that is relative to the box. Let the lower ball move at the speed of say 70 percent of the speed of light to the right relative to the box. Then it seems that the lower ball is moving relative to the ground at 140 percent of the speed of light?

Let us estimate the masses of the balls. If we stand inside the box, the masses of both balls are equal and because of Eq. (2.6) they are larger than their resting-masses by the factor $1/\gamma$ belonging to the speed of 70 percent of c. Looking at it from outside, these two masses add up to the **resting-mass** of the box because we assumed the box itself to have nearly no mass.

Now look at these masses while standing still on the ground. The box moves at the same speed to the right as the lower ball inside the box, so its moving mass is again larger by the inverse gamma factor $1/\gamma$. Hence the total mass of the box is twice the resting-mass of one ball, divided by γ^2.

Fig. 2.13 Masses move inside a light moving box

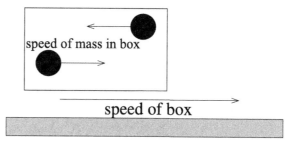

Next look from the ground at each ball separately. The right ball moves at the same speed as the box, but to the left, so it rests relative to the ground. Hence its mass is now just its resting-mass. What is the mass of the left ball? The left ball has just such more mass as the right ball has less because their masses should add up to the same total mass of the box as before. However, we know that the higher the speed is the more mass increases. Hence the resultant speed of the left-hand ball, relative to the ground, must be less than twice the speed with which the balls are moving relative to the box. In fact, it is so much slower that again it is not faster than light! In other words:

Relativistic speed addition
If some box moves relative to the ground at some speed and some mass moves inside that box at some speed in the same direction, then that mass moves relative to the ground at a speed which is *so much less* than the sum of the speeds so that it never can be faster than light.

For the actual calculation, see Appendix A.4. The result is in terms of fractions of the speed of light

$$\frac{\text{total speed}}{c} = \frac{\dfrac{\text{speed of box}}{c} + \dfrac{\text{speed of mass in box}}{c}}{1 + \dfrac{\text{speed of box}}{c} \times \dfrac{\text{speed of mass in box}}{c}} \tag{2.8}$$

For our example we get speeds

$$\frac{\text{total speed}}{c} = \frac{0.7 + 0.7}{1 + 0.7 \times 0.7} \approx 0.94$$

which is of course less than the speed of light because it *must come out* that way!

Chapter 3
Light, Electricity, and Magnetism

3.1 Electric Charge and Speed

We have seen that the theory of relativity provides us with a wealth of new insight about the relations between light, time, space, mass, energy, and other quantities. Some quantities were earlier supposed to be invariable. For example, before the theory of relativity, the total mass involved in some experiment was supposed not to change or to be **conserved**, as physicists refer to it. Another example is the conservation of energy which we discussed in Sect. 1.9. Relativity merged energy and mass, as Einstein put it in 1906:[1]

> Nach der in dieser Arbeit entwickelten Auffassung ist der Satz von der Konstanz der Masse ein Spezialfall des Energieprinzipes.

That is to say:

> Mass in the capacity of being energy is conserved as energy.

It was certainly hard to reexamine such basic concepts as space and time from the very start: things that the philosopher Kant thought to be contained in our brain to be able to grasp all the world. However, there *was* a road map at hand, a blueprint of what was to be expected: that was and still is the theory of moving electrical charges, that is **electrodynamics**. The reason is simple:

---

[1] A. Einstein. Das Prinzip von der Erhaltung der Schwerpunktsbewegung und die Trägheit der Energie. Annalen der Physik, volume 20, page 627, 1906.

© Springer International Publishing Switzerland 2015
K. Fischer, *Relativity for Everyone*, Undergraduate Lecture Notes in Physics, DOI 10.1007/978-3-319-17891-2_3

While mass, time, and length change at speeds relative to the observer, **electrical charge does not change** at all!

There is even more. In Sect. 2.2, we got the impression that relativistic effects are important only at large speeds because the γ factor differs from one only less than a part in a million-million for ordinary speed on Earth. This is incorrect!

We will see in this chapter that the theory of relativity enables us to explain *magnetism*. It is caused by electric charges which move much less than a millimeter per second, obeying the laws of relativity. In other words, magnetism is relativity that is visible at really *low* speed.

3.2 Electric Charges and Magnets

Maxwell created the theory of electromagnetism in the nineteenth century. Before that, electricity and magnetism were considered to be separate things. However, then people observed things such as

An **electric current** in a wire makes a nearby **magnetic** compass needle move.

and started to take a closer look which yielded:

In an **electric motor** electric current through a wire makes magnets move relative to those wires. Conversely, in an **electric generator**, magnets moving relative to wires create electric current.

We sketched the principle of the electric motor in Fig. 3.1. Electric charges that are in this case electrons move through the black wire loop. In other words: an *electric current* runs through the wire. The magnetic field exerts a force on these moving electrons.

This force is called the **Lorentz force**. We get the direction of the force from the **left-hand rule** [2] which we explain below Fig. 3.2.

---

[2]Some textbooks specify the electrical current to flow in *opposite direction* to the flow of the electrons. Then the left-hand rule becomes a right-hand rule, but of course the physical phenomena do not change.

Some other textbooks use the *Fleming left-hand rule*. However, this rule uses the fingers in a different order, and would become a right-hand rule in our case. The electrons move in the direction of the right middle finger, and the magnetic field points in the direction of the right forefinger, so that the force on the moving electrons points in the direction of the thumb, that is, again upwards.

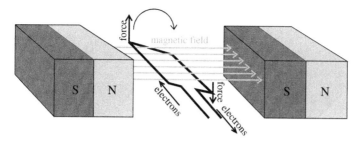

Fig. 3.1 The magnetic field of magnets exerts a force on electrical charges moving relative to it. The magnetic field itself is invisible. This is the basic model for the electric motor

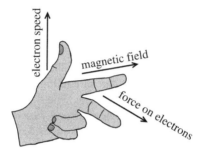

Fig. 3.2 Left-hand rule for negative charges. Spread the thumb, the *index finger*, and the *middle finger* of your *left* hand, such that they stand at *right* angles to each other. Then turn the *left* hand such that the negative charge, electrons in our case, move in the direction of the thumb and that the magnetic field points in the direction of the *index finger*. Then there will be a force acting on the moving electrons in the direction of the middle finger. Please try this rule now for the situation shown in Fig. 3.1 by turning your *left* hand accordingly!

Let us see how the Lorentz force acts for our model of the electric motor in Fig. 3.1. The electrons enter the magnetic field and move to the rear, and the magnetic field points from the north pole of the left magnet to the south pole of the right magnet, that is to the right. Hence, according to the left-hand rule a force is pushing the electrons *upwards*. Therefore this force pushes the left part of the wire loop upwards.

Then the electrons turn back behind the magnetic field and move to the front. The left-hand rule tells us that there is a force pushing the electrons, and therefore the right part of the wire loop, *downwards*. In total, we see that the wire loop will begin to rotate clockwise. This thought experiment shows how an **electric motor** works.

In Fig. 3.3, we see the opposite effect at work: now no current is flowing through the black wire. *We* turn the wire loop clockwise. Then the electrons being in the left part of the wire move upwards. The left-hand rule tells us that there is a force pushing the electrons to the *front*. Likewise, the electrons in the right part of the wire move downwards. The left-hand rule tells us that a force is pushing these electrons to the *rear*. Hence an electric current will begin to flow through the wire. This thought experiment shows how an **electric generator** works.

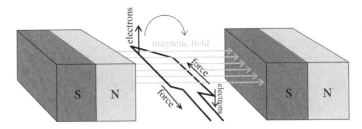

Fig. 3.3 Again, the magnetic field of magnets exerts a force on electrical charges moving relative to it. This time the same setup as before serves as the basic model for the electric generator

3.3 Electric and Magnetic Fields

What is more, by carefully observing and interpreting the experiments, Maxwell concluded that the energy of a magnet or electric charge spreads out in space as an **electric field** or **magnetic field**. Indeed, the wire in Figs. 3.1 and 3.3 does not touch the magnet but acts or reacts to the invisible magnetic field of the magnet.

If an electric field varies with time, it creates around it a certain amount of magnetic field, and vice versa. He summarized his findings in his **Maxwell equations**.

Actually, the Maxwell equations *alone* do not tell us how a mass carrying an electric charge *reacts* to an electric or magnetic field. This law is logically independent of the Maxwell equations. The force which an electromagnetic field is exerting on a charged mass is called the **Lorentz force**. We saw the Lorentz force in action in Figs. 3.1, 3.2, and 3.3. The Maxwell equations together with the Lorentz force make up **electrodynamics**.

After convincing himself that his equations described the experiments correctly, Maxwell pushed his new theory to an extreme. Even if there are *no* electric charges or magnets around in empty space, suitably varying electric and magnetic fields can sustain themselves. However, they *cannot rest* at the same place but must travel at a certain fixed speed. Maxwell could even calculate this speed from his equations and found that it was the same as the speed of light! Hence light is nothing but a wave of varying electric and magnetic fields, called an **electromagnetic wave**.

Maxwell's great achievement was not only to unify electric and magnetic phenomena under one roof but also to show that light itself is an electromagnetic phenomenon. Therefore he actually unified *three* types of phenomena: electric, magnetic, and optic.

What is equally important: the absolute, fixed speed of light comes out of the Maxwell equations as a constant of nature *without further ado*. We already mentioned that electrical charge also does not depend on the speed of the charge and therefore it is absolute. In short, electrodynamics fits in with the theory of relativity *from the start*.

Therefore we can use electrodynamic phenomena as a road map to build the theory of relativity. We do so by using a thought experiment: an electrical current running through a wire and creating a magnetic field. In fact, this thought experiment was the starting point for Einstein's article about the theory of special relativity. Its title is **"On the Electrodynamics of Moving Bodies"**, and now you understand why!

3.4 Magnetic Field from Electric Current

In Fig. 3.4 we sketched part of a long, straight wire made of some metal. The wire itself rests relative to us. Through the wire a steady electric current flows, namely electrons from the left to the right. For our thought experiment, we choose a wire of very low electrical resistivity, created for example by freezing the wire to a very low temperature so that the atoms nearly freeze out and do not disturb the moving electrons. We assume from now on that there is no electrical voltage needed in the wire to sustain the flow of the electrons.

By the way, the electrons in a typical metal wire are actually slow. They move at less than a tenth of a millimeter per second!

In Fig. 3.4 we sketched the electrons of the electrical current as white ovals with a black minus-sign. Some of the atoms of the metal supply the moving electrons in the wire so that they now lack an electron. The charge of these atoms is therefore now positive. We sketched these atoms as black plus-signs. They rest in the wire. The total wire is electrically neutral, that is, it carries no *net* charge.

We can check that by putting a negative charge *resting* in front of the wire such as one we sketched as black oval with a white minus-sign. Nothing happens. The black charge will not begin to move.

However, the electric current in the wire produces a magnetic field around the wire, the direction of which we sketched by the lines going around the wire. Then the **left-hand rule** of Fig. 3.2 tells us that the magnetic field *attracts* the black electrical charge if that charge *moves relative* to the field *to the right*, as in Fig. 3.5. The experiment tells us that this **Lorentz force** grows in proportion to the speed of the black charge. The experiment tells us also that this force grows in proportion to the current as well, that is in proportion to the speed of the electrons. Therefore, *if* the black negative charge is moving at the *same* speed as the electrons to the right, then the magnetic field of the wire attracts the negative black charge with a force growing in proportion to the *square* of this speed.

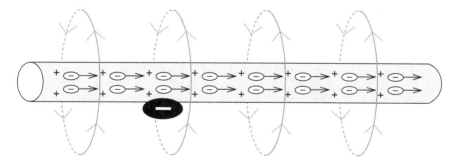

Fig. 3.4 A steady current of electrons moves in a straight, long wire from the *left* to the *right*. The negative, *black* charge with the *white* minus-sign rests in front of the wire, and remains there

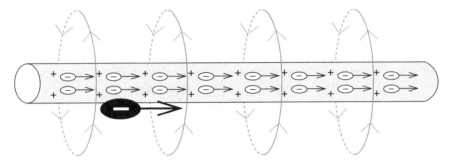

Fig. 3.5 The negative *black* charge moves to the *right*. Then the magnetic field of the wire attracts this charge

3.4.1 The Faraday Paradox

Next, we are moving to the right at the same speed as the black negative charge. Hence, for us the electrons of the current in the wire as well as the black negative charge outside are now *resting*. However, the positively charged atoms move at the same amount of speed to the *left*. In other words: the wire itself moves to the left, as sketched in Fig. 3.6.

Let us at first only use the **first law of relativity**. The physical effects do not depend on whether the observer moves at a steady speed in one direction. We see now a current of *positive* charges of the same size moving at the same speed to the *left*, as we saw before moving to the *right* as *negative* charges. Hence these positively moving charges produce the same magnetic field. Thus we have the same situation as before. In particular, we are again *resting* relative to the magnetic field!

In other words: in the situation of Fig. 3.5, we rested relative to the wire and relative to the magnetic field. However, in the situation of Fig. 3.6, we move along the wire but still rest relative to the same magnetic field as before! This raises the question: is this "magnetic field" we drew around the wire, a real, physical quantity?

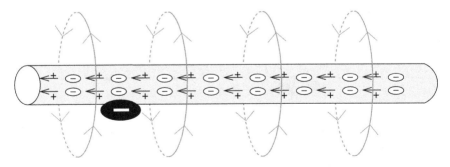

Fig. 3.6 Now the positive charges of the wire move to the *left*. We and the outside negative charge rest again with the magnetic field

After all, it is invisible, and seems to rest even if we move along the wire! This is called the **Faraday paradox**.[3]

In order to check the magnetic field, we now look at the force it exerts on the black negative charge.

3.4.2 No Attraction Without Relativity

Let us again at first only use the first law of relativity: in Fig. 3.5, the outside negative charge moves relative to the magnetic field, so that the magnetic field attracts it. However, in Fig. 3.6, the outside negative charge does *not* move relative to the magnetic field, so the magnetic field will *not* attract it! This is weird, and exactly this was the situation when Einstein published his article about the "electrodynamics of moving bodies".

3.4.3 Attraction with Relativity

Next we also use a consequence of the **second law of relativity**, that is to say the relativity of lengths in the direction of speed of Sect. 2.4.1. We know that the resting wire was electrically neutral. This means that per meter, say, the number of positive charges on the atoms and the number of negative charges were the same. As moving observer we see now:

1. The charge *per* electron or atom does *not* change. That is an experimental fact suggesting already that relativity and electrodynamics fit.
2. Suppose that we put a rod parallel to the wire, resting with it. Because we now move relative to this rod, the rod looks *shorter* for us, so the same number of positive charges fits into a shorter distance. Hence per meter, relative to us, there are now *more* positive charges than before.
3. Suppose that a rod moved at the same speed as the electrons to the right. Because we now rest relative to this rod, it looks *longer* for us so that the same number of negative charges fits into a longer distance. Hence per meter, relative to us, there are now *less* negative charges of the electrons than before.

Altogether the wire has now per meter a net positive charge. We sketched this in Fig. 3.7, exaggerating it for a better view.

We also know from Eq. (2.4) that the γ factor for slow motion differs from one by an amount which is in proportion to the square of the speed. Hence the net positive charge on the wire is also in that proportion. Therefore now the positively charged wire *attracts* the outer negative charge because that is what charges of the opposite

---

[3]Many textbooks use a rotating magnet for this thought experiment.

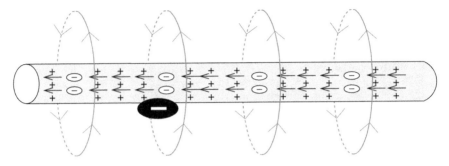

Fig. 3.7 The moving wire has now a net positive charge and therefore attracts the negative charge in front of it

sign do in proportion to their charges. Hence the wire attracts the positive charge with a force in proportion to the square of the speed.

Compare with the resting wire of Fig. 3.5: in both cases the force attracts and in both cases the force grows in proportion to the square of the speed! Hence we can already guess without any calculation that the two situations are identical.

We see that with the help of the theory of relativity we can understand electrodynamics: we just *explained* the magnetic Lorentz force and the left-hand rule! Magnetic fields are nothing but electric fields from moving charges acting upon moving charges.

Without knowing quantum theory, we can already guess that in a permanent magnet there must be moving charges to sustain the magnetic field!

When estimating the γ factor, we assumed that the electrons or atoms only move slowly but that is not necessary. From both points of view of the observer, the wire attracts the outside negative charge in *exactly* the same amount.

What about the Faraday paradox? Since there are in Fig. 3.7 more positive charges moving at the same speed to the left than there are in Fig. 3.5 with negative charges moving to the right, the electric current is now stronger. Hence the magnetic field is *not* the same as before. It has *increased*!

We have seen in this chapter that relativity is *not* only about large speeds and spaceships. It is also about *very slow* bodies—for example, the electrons in a wire move with less than a millimeter per second! And it is about everyday electrical appliances. Magnetism itself *is* a consequence of the theory of relativity.

Chapter 4
Acceleration and Inertia

Let us come back to the theme of the first chapters: sending light at different speeds. Up to now, the sender and observer moved steadily at a speed keeping both magnitude and direction constant. The part of the theory of relativity dealing with these phenomena is the **theory of special relativity**. However, to arrive at some speed, or to change the direction, we must **accelerate**. We ask: how does acceleration of a body affect time, length, mass, or energy? This will eventually help us understanding gravity.

4.1 Rotating Motion: Twin Paradox 1

We begin with the simplest case in which speed changes its direction but not its magnitude, as on the merry-go-round in Fig. 4.1. The merry-go-round has the form of a disk with a hole in its center. We placed the merry-go-round on such a small planet, that we can neglect its gravity.

Suppose we ride at the rim of the rotating merry-go-round. The merry-go-round pulls us constantly inwards, along a circle. As it is, our speed is always changing *relative* to the ground, that is, it changes its direction: we **accelerate by rotation**. We feel our inertia because mass itself resists acceleration. If we slip off the merry-go-round, then we would move along a *straight line*, as our inertial mass prefers to do. We would change into an **inertial state** in which we continue to move steadily at a speed keeping both magnitude and direction constant.

What happens if we stand with a clock on the ground at the center of the merry-go-round? Then we do not turn around. Nothing accelerates us. We are in an inertial state. For this state, we know by *experience* that time passes steadily. We see a friend with his clock riding on the rim of the merry-go-round at some speed. For a very short time interval, we can think that our friend at the rim is going nearly straight ahead. We use our knowledge about moving clocks and conclude that the clock at the rim will *run more slowly*, and that by the γ factor of Eq. 2.3.

© Springer International Publishing Switzerland 2015
K. Fischer, *Relativity for Everyone*, Undergraduate Lecture Notes
in Physics, DOI 10.1007/978-3-319-17891-2_4

Fig. 4.1 Measuring time on a merry-go-round

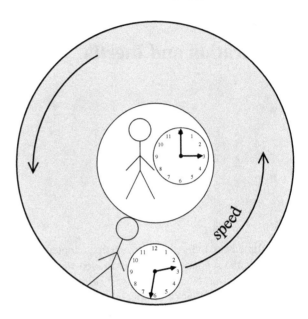

Because the clock at the rim does not move away from us, we can wait as long as we wish and we see the clock at the rim slowing down against our clock in the center more and more. This leads to a weird thought experiment. Suppose that both observers were identical *twins*. At first, they both rest in the center of the merry-go-round. Then one of them steps on the merry-go-round and moves to the rim. He rides there some time and then returns back to the center. Certainly his clock will react somehow during the time in which he moves towards the rim, and afterwards when he returns. However, that does not depend on what happens while he is riding on the rim: he can do this as long as he wishes. In the end, when he has returned after riding on the rim for a long time, he is *younger* than the twin who waited all the time in the center!

This is the famous **twin paradox** or **clock paradox**. In fact, it is no paradox but measured reality!

For example, in 1977 **muons** were sent along a circular tube of a 14-m diameter at the European Organization for Nuclear Research **CERN** in Geneva. Muons with negative electric charge are a kind of heavy version of an electron. They decay into electrons and other particles. A muon *resting* near us has a lifetime of about $2 \cdot 10^{-6}$ seconds.

However, "on the merry-go-round", that is inside the circular tube, the muons circled at a speed of $0.99942 \cdot c$. Hence the γ-factor is about $\frac{1}{29}$ so that they should live about 29 times *longer* than when resting near us. This is precisely what was found![1]

---

[1] You can download the original paper here: http://cds.cern.ch/record/929453/files/ep63_001.pdf. Their γ is our γ^{-1}.

4.2 Rotating Motion: Not School Geometry

Next, our friend at the rim wants to check out how lengths are effected by the rotation. Standing in the center we produce many small rods and give them to our friend at the rim. The rods are so small that by joining them he can estimate the length of the rim quite precisely.

We know from Sect. 2.4.2 that lengths at right angles to speed do not change. Therefore we agree with our friend that the diameter of the merry-go-round does not change as it turns around.

Then our friend measures the length of the rim of the merry-go-round, as in Fig. 4.2. A small part of the rim appears nearly straight, if we only zoom in enough, as in the right part of the figure. Hence for small enough rods, our friend can estimate the length of the rim in terms of the number of rods needed to circle the rim.

Suppose at first that the merry-go-round is at rest. Our friend sends a light beam along the rod at his feet. The light beam passes his feet at the speed c. This speed is the length of the rod divided by the "very short time" needed for the light to pass along the rod. Therefore the length of the rod is c times this "very short time". He reads this "very short time" from a clock that he is carrying so that he knows the length of the rod. After counting the number of rods that exhaust the rim, he will conclude that the length of the rim is π times the length of the diameter, as we expect from **school geometry**.

Next, as for the rotating merry-go-round, the light beam will still pass along the rod at his feet because the rod lies in the direction of the rotation speed. The light beam passes at the rod still at the same speed c because the speed of light is absolute and for the "very short time" our friend travels nearly straight ahead, along the rim.

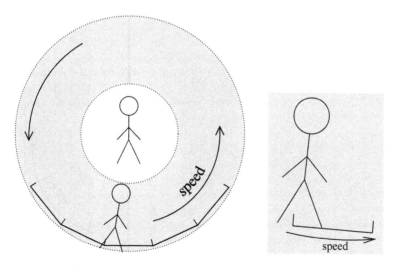

Fig. 4.2 Measuring lengths on a merry-go-round. Short lengths along the rim look nearly straight, as we see in the closeup on the *right*

He reads the "very short time" from his clock and finds that the light beam has not made it along the rod but traveled only along some *part* of it! Why is that? For our rotating friend time is passing *more slowly* than for us resting in the center, that is in shorter intervals and shorter by the γ factor of the rotation speed. Hence the light beam will not make it during the "very short time" which he reads from the clock. However, for our friend the length of the rod is still the speed of light c times the time needed the light beam to travel along the rod. In other words, for him the rod has become *longer*.

However, there are still the same number of rods circling the rim. Therefore, for him the *ratio* between the length of the rim and the diameter is *larger* than π! To be precise: the ratio is by a factor $1/\gamma$ larger. Hence our friend decides to shorten the rods by the factor γ. Then he needs now *more* rods to circle the rim of the disk. In Fig. 4.3, we overdid the effect for a better view.

Up to now, when bodies moved at a constant speed straight ahead, we saw that time and length among others changed relative to the speed of the observer. However, at least we still could use **school geometry** or what mathematicians call **Euclidean geometry**. For example, we calculated the γ factor using the **Pythagoras theorem** in Sect. 2.2 which is part of the Euclidean geometry. This geometry is a rigid system of theorems. If one of them ceases to be true, many others will fail as well. One of these theorems is that the ratio of the length of the perimeter and the diameter of a circle is always π. In our case of the rotating merry-go-round, we already saw that this is not any longer true. The rim of a rotating merry-go-round is longer than its diameter times π! We can only conclude: if we are under **acceleration**, the **school geometry is no longer valid**!

Fig. 4.3 The rim is longer than π times the diameter, when measured as number of rods

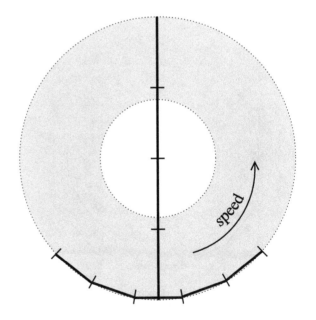

4.3 Straight Motion and Acceleration

The weirdest things happen under acceleration. This is not limited to rotating motion. Let us now put the light sender to the left of a transparent elevator, as in Fig. 4.4. The upwards pointing arrow shows that the elevator is **accelerating upwards**. We stand outside and see the elevator passing by. At this instant, we send a light beam through point A.

We see this beam passing through the glass walls of the elevator, in a horizontal, straight line, from point A to the point B, as in Fig. 4.5. This is the fastest way some body can travel from A to B because nothing is faster than light.

What does our friend in the elevator see? Please have a look at Fig. 4.6. It shows the moment when the light is leaving the elevator. As the light passes through the elevator from point A on, the elevator gathers speed upwards so that for our friend the light *bends* downwards. He sees the light beam leaving the elevator at some point B further downwards.

We drew a straight, dotted line on the wall of the elevator between the points A and B. Is this path not *shorter* than the path the light has chosen? If so, is our friend not able to send some signal along that shorter, straight path, *faster than light*?

The answer is: the question was wrong! We drew the straight, dotted line *before* the elevator began to accelerate. For us time proceeds then at an even pace. Once the

Fig. 4.4 We send a light beam through an accelerating, transparent elevator

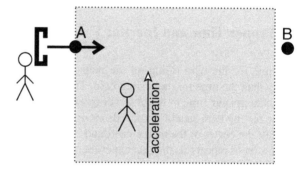

Fig. 4.5 Outside we are in an inertial state so that light travels along a *straight line*

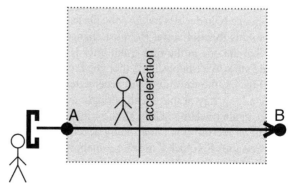

Fig. 4.6 Light bends for the observer under acceleration

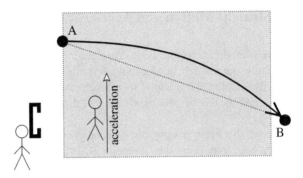

elevator is accelerating, time inside the elevator slows down more and more. Straight lines will deform because vertical lengths will shrink and horizontal lengths will remain the same. Further, our friend *cannot* even measure "the" length of the dotted line. Indeed, while he is measuring, time and lengths will distort even more! He is in **space-time**, not only in space!

We see that the easiest way to handle rods and clocks is the one in an inertial state: time steadily proceeds and lengths do not shrink. Once you begin to accelerate, lengths can change as time proceeds. The Euclidean geometry is no longer valid so that things get complicated.

4.4 Proper Time and Inertia: Twin Paradox 2

The time for the twin riding on the merry-go-round in Sect. 4.1 proceeded more slowly than the time for the resting twin. This time measured by one's own clock is called the **proper time** of the observer or a body. Hence we can say: the proper time for the resting twin steadily proceeds *more quickly* than the proper time of the twin who left the center of the merry-go-round.

Now let us repeat this thought experiment in another way, as shown in Fig. 4.7. At first, both twins rest near each other at the start, not feeling any acceleration. Their proper time steadily proceeds.

Then one twin decides to *leave* the place. Hence he has not only to leave the *place* but also this **inertial state**. He must change his speed to get away. He travels on a route like the one in the figure and after he has returned, he finds that his clock is running slow. We can push this thought experiment to its extreme:

In Fig. 4.8, the traveling twin accelerates at first to a certain speed, and then he travels a long way at this speed, straight away, in an inertial state. Then he gently returns, again traveling nearly all the way home at the same constant speed straight away in an inertial state, and finally accelerates to stop at the start beside the first twin. Certainly his clock will react somehow during his initial and final acceleration, or when he returns. However, that does not depend on what happens when he is

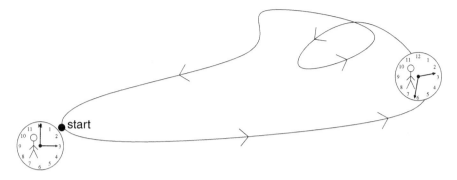

Fig. 4.7 No matter what course the twin takes. After returning home, he finds that his clock is running slow

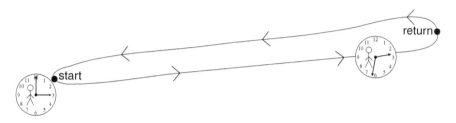

Fig. 4.8 The clock of the twin on the left remaining at the start remains in the same inertial state and therefore is running more quickly

traveling at a constant speed back and forth. The longer the distance to the point of return, the longer he travels straightly at a constant speed back and forth, and the more his clock slows down.

Therefore we can neglect these accelerations. In comparison to the time of the twin remaining at the start, the time of the traveling twin advances less by the γ-factor of his mainly constant speed.

However, is it really during this long trip in the inertial state that the traveling twin's clock is going more slowly? He may say: "I only saw my twin brother going away from *me* to the left. I certainly was in an inertial state for a long time during which I saw my twin brother moving away from me at a constant speed, straight away and later approaching me, again at a constant speed. So, during that time my inertial state is as good as his and from what I learned in Chap. 2, I reckon that *his* clock slowed down, not mine!"

This is again the **clock paradox** or **twin paradox**.

However, his opinion is wrong: we compare the clocks *after the traveling twin returned and is again resting with the other twin*. It is *irrelevant* what he *thinks* about what would happen in the meantime because that was not the question but only which time he had accumulated in the end.

In fact, *even* the traveling twin *can observe* that his clock does only advance the fraction γ of the remaining twin's clock! Why is that? Let us think about the *distance* between the start and return points in Fig. 4.8. The twin remaining at the start measures this distance, for example by sending a light beam to the return point where a mirror reflects the light back to the start point. Based on the time during which the light was traveling, he calculates the distance. From this distance and the speed of the traveling twin, he calculates the time that the traveling twin needs to return to the start. Of course, that is the time that runs *for the twin remaining at the start*.

Now, what distance will the traveling twin measure? We already saw in Sect. 2.4.1 that distances in the direction of speed will *shrink* by the γ-factor of that speed. Therefore the distance between start and return shrinks for the traveling twin by that γ-factor. So he indeed needs only the fraction γ of the time which the twin observes at the start, because he has to travel *a lesser distance* than what the remaining twin measured.

In short: for the twin remaining at the start, the *time* of the traveling twin is passing more slowly by the factor γ, while for the traveling twin, the *distance* is shorter by this factor γ. So there is no paradox: it all fits!

4.5 Inertial State, Acceleration, and Proper Time

Let us sum up. The crucial difference between the two twins in the thought experiment in Sect. 4.4 is that the traveling twin *changed* his inertial state. Similarly, the twin in Sect. 4.1 riding on the merry-go-round changes his inertial state constantly because the direction of his speed is constantly changing. Changing the inertial state means changing the amount or the direction of speed, that is, *accelerating*. However, we saw that in both cases it was *not* the *amount of acceleration* but the *amount of the speed* that determined how much the clocks of the two twins differ. If, for example, the merry-go-round is only half the diameter and its rim turns at the same speed, the twin riding at the rim will feel a larger acceleration, but his time will slow down by the same proportion.

In any case, the time of the twin who changed his inertial state was running more slowly against the time of the twin who stayed in the same inertial state. Turning the tables, this means that as long as we drift freely in space, in the same inertial state, our proper time evolves always *the fastest*. To coin a catch-phrase:

Because matter tends to remain inert, it moves without force acting upon it so that it experiences the **longest proper time possible**.

Chapter 5
Inertia and Gravity

Matter drifting in empty space resists being accelerated: it has inertia. However, near large masses like the Earth, for example, a smaller mass seems to accelerate towards the large mass. Starting from this simple observation will help us to understand better how to deal with acceleration and gravity.

We feel gravity everyday, beginning when we get up in the morning, when we go upstairs or downstairs, or drop a cup, or such like. It is too common to be astonishing. And yet gravity is one of the most mysterious things in the universe, if we think a little bit about it. For this thought experiment we need pure gravity, not marred by air drag. So let us make a thought experiment, similar to what high school teachers usually do. We put a bird feather and a heavy metal ball into a glass tube and close the tube at both ends with some plugs. Through one of the plugs we drive a thin tube connected to a vacuum pump, pump the air out of the glass tube, and seal it. Then we hold the air-evacuated tube vertically and invert it quickly. The result will be as shown in Fig. 5.1.

The ball and the feather fall down *equally quickly*! On Earth we have to get rid of the air to see this, but on the moon the astronaut David Scott from the Apollo 15 mission did exactly that. He dropped a hammer from his right hand and a feather from his left hand at the same time, and both reached the ground at the same time. Just search for the keywords "hammer feather Apollo" on the internet, and you even can see a video from the original television broadcast.

Over the last 100 years or so, physicists repeated this kind of experiment with more and more accuracy and with all kinds of materials including even subatomic particles, and they always found this result:

All matter *reacts* in the same way to the same gravity.

This is astonishing because it shows that inertia and gravity are somehow connected. Let us sort it out. You can measure the inertia of a stone of say 1 kg on an ice skating rink by pushing the mass, as we already saw in Fig. 1.10. This has nothing

© Springer International Publishing Switzerland 2015

K. Fischer, *Relativity for Everyone*, Undergraduate Lecture Notes
in Physics, DOI 10.1007/978-3-319-17891-2_5

Fig. 5.1 A bird feather falls
as fast as a metal ball inside
an air-evacuated tube

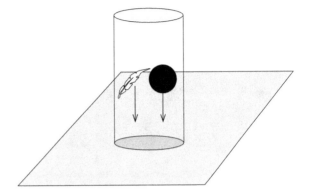

to do with gravity. You can do the same in empty space, away from Earth, and mass
will still resist being pushed.

Or we can *weigh* the stone. Again you will find "1 kg", this time as weight or
gravitational mass. Here the stone on the scale is at rest. So it does not resist any
acceleration, it *rather tends* to fall towards the center of Earth.

The point is now when you take a 1-kg piece of some material and another 2-kg
piece of another material as measured on the ice skating rink, then the first material
will *weigh* exactly 1 kg and the second will *weigh* exactly 2 kg. In other words, their
inertial mass and their gravitational mass are *exactly* equal. Why? If the second
material weighed say 2.1 kg, then in the glass tube it would fall *faster* than the first
material. Nobody has seen such a thing up to now.

> **Inertial and gravitational mass** are exactly equal for all materials.

5.1 Gravity Is Not a Force

So what exactly happens when the feather and the ball fall down the tube? The
gravitational mass of the ball tries to accelerate the ball downwards. At the same
time, the equally large inertial mass resists this acceleration. The same goes for the
feather. Here is now the point: because inertial and gravitational mass are exactly
equal, the ball *does not accelerate at all*! The ball is **free-falling** and is therefore in
an inertial state.

However, if you are standing nearby you may argue: but I *see* the ball accelerating
towards the Earth! Answer: no, *you* are accelerating away from Earth! While standing
on the ground, you feel that upward **acceleration** and therefore your imagination is
hampered by all kinds of weird things that can happen if we are in an accelerated
state, as described for the merry-go-round and the elevator in Chap. 4.

Fig. 5.2 Suddenly someone
pulls the elevator with the
rope

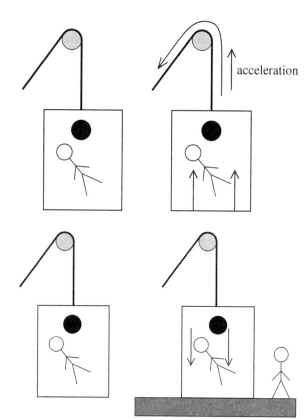

acceleration

Fig. 5.3 Suddenly a planet
appears under the elevator

So in order to really sort it out, we had better remain in an inertial state. To clarify
things, we make a number of thought experiments by using *elevators*, in a similar
way as Einstein originally did. Please have a look at Fig. 5.2.

In the picture on the left, we sketched an elevator floating in empty space, far away
from any planet, star or other large mass, and no air inside. The astronaut wearing
a space suite is inside the elevator and the ball nearby floats freely as well. Then
suddenly someone pulls the elevator so that, in the picture on the right, the astronaut
sees the floor of the elevator accelerating up towards him. However, the astronaut as
well as the ball *are still floating freely*. No force is pulling them! Hence he reckons
that someone is pulling the rope and accelerating the elevator upwards.

Or does he? What if in reality it all happens as in Fig. 5.3? This time nobody pulls
the rope but suddenly a planet sneaked under the elevator. Still the astronaut does
not feel *any acceleration* and the nearby ball does not move relative to him. This is
because the inertial and gravitational mass of the astronaut and the ball are exactly
equal. He and the ball are still in an inertial state, free-falling! He again sees the floor
of the elevator accelerating towards *him* and the ball. However, the observer standing
on the surface of the planet sees the astronaut and the ball falling downwards.

By measuring acceleration *alone* the astronaut in the elevator cannot decide from inside the elevator which of the two cases is true. He has to *look outside* the elevator to find out.

This is the famous *equivalence principle* by Einstein:

Equivalence Principle
By measuring our acceleration alone, we can in no way decide whether gravity acts upon us. In particular, **free-falling** masses are in an **inertial state**, as they would be floating freely in empty space.

The equivalence principle will be our handy tool to understand gravity. Because free-falling masses are in an inertial state, their proper time is proceeding steadily. We can use our previously gathered knowledge about this state to explore the surroundings of large masses, i.e. to find out how gravity works. The equivalence principle links inertia with gravity.

Let us start by again considering the merry-go-round of Sect. 4.1. For an observer riding on the rim of the merry-go-round, Euclidean geometry does not work. His clock is running more slowly than the clock of someone resting outside in an inertial state.

We saw this was the case because he is constantly changing the direction of its speed relative to the center of the merry-go-round. Now compare with what happens on the surface of Earth. We being on the surface accelerate constantly upwards, that is away from the center of Earth. In other words: we are not in an inertial state. Therefore we can expect that as for us under gravity, Euclidean geometry does not work either and that time will run more slowly than for an observer resting far away from Earth.

However, we know from Sect. 4.5 that the amount of acceleration that we feel while riding on the merry-go-round *does not determine* how much clocks slow down. The equivalence principle tells us that therefore the amount of acceleration while standing on Earth *does not determine* how much clocks slow down under gravity!

Therefore we have to analyze now in more detail how gravity influences space and time.

5.2 Gravity Bends Space-Time

Image a rocket going around Earth, with its engine switched off. The rocket is free-falling, not towards Earth but around it, as we sketched in Fig. 5.4. We place ourselves resting far enough away from Earth so that we are nearly in an inertial state.[1] By the

---

[1] For the meaning of "far enough" or "nearly", see the last section of the preface.

Fig. 5.4 Rocket free-falling
around Earth

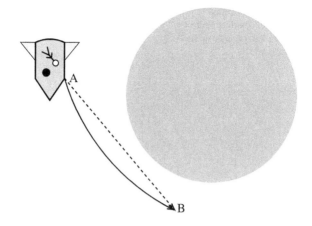

equivalence principle, the rocket is in an inertial state as we are. However, we see that the rocket does not travel along the dashed straight line. What forces the rocket around the Earth along the bended path, drawn as a solid line?

5.2.1 Bended Surface

To find the answer to that question, think of an analogy: choose two points *A* and *B*, somewhere in empty space, far away from any planet or star or any other large mass. Try to find the shortest path between these points. Clearly this is the straight path connecting them. How can we trace this path? Start at *A*, head towards *B*, and keep going *straight ahead*.

Next, suppose the two points *A* and *B* are along the equator in the Pacific Ocean, say *A* being west of *B*. Try again to find the shortest path between the two points. The shortest path between the two points is a tunnel along the straight dashed line connecting them, as sketched in Fig. 5.5.

If we restrict ourselves to the Earth's surface, the shortest path is the solid line which is part of the equator: it *bends*. How can we trace this path? Start at *A* and keep *straight ahead*, eastwards.

For short distances, the ocean's surface looks *flat*. For short distances, keeping straight ahead to the horizon perhaps means to follow a straight path. Hence the shortest path connecting two points on a bended surface is not straight, but the **straightest path**.

Such shortest paths are always the straightest paths but on a bended surface only the "short enough" straightest paths are necessarily the shortest paths: let us again start at *A* keeping *straight ahead*, moving *westwards* along the equator, following the fat line in Fig. 5.6. This path is nearly a straight line for short distances and up to some intermediate point *C* being not too far away, it is the shortest of all possible

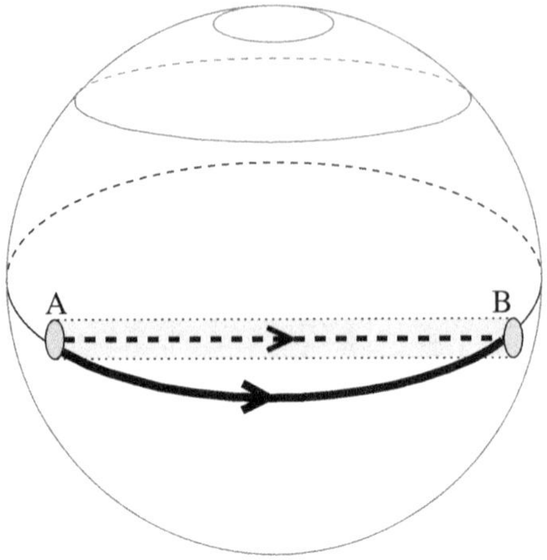

Fig. 5.5 On a bended
surface, the shortest path
between two points usually
bends as well

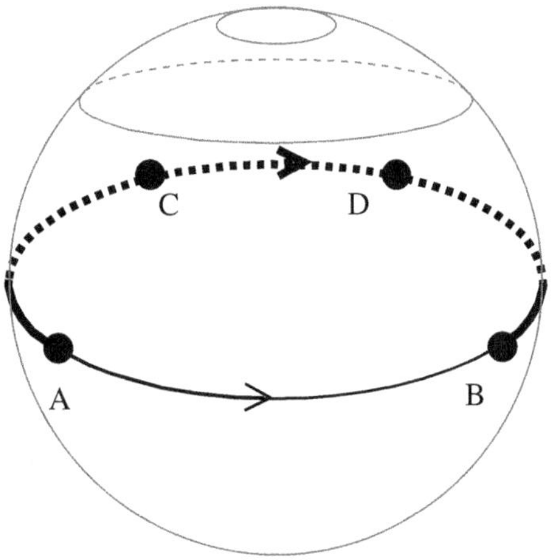

Fig. 5.6 On a bended
surface, the straightest path
between two points is
sometimes not the shortest
path of all

paths connecting A and C. You can see this in Fig. 5.7 by comparing it for example
with the gray path connecting A and C. If we follow this straightest path further on,
it is again the shortest path between the points C and D, and afterwards between the
point D and the endpoint B. Hence this straightest path is the shortest of all paths
which cross the total gray path at "short enough" intervals, see Fig. 5.7.

Hence the total path is the straightest path from point A to point B, if we start in
a westward direction. However, this path is of course longer than the eastward path
in Fig. 5.5. We will encounter such a straightest path in space-time in Sect. 6.2.

Such a straightest path is the natural analog to the straight line in a plane. Its name
is **geodesic**. The name tells: *geodesy* is the science of measuring and mapping the
surface of Earth...

Fig. 5.7 The straightest path is shorter than all **nearby paths**, that is those which cross it again and again at short enough intervals

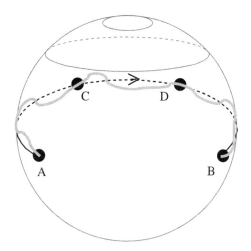

5.2.2 Bended Space-Time

Over short distances the free-falling rocket travels nearly along a straight path at a constant speed. For longer distances the path can bend and the speed can change, as we have seen. However, this path from A to B in space is usually *not* the straightest path in space, as it is clear from Fig. 5.4. Then what does gravity bend?

> Gravity does not only bend space, gravity **bends** space *and* time, called **space-time**.

Let us see how.

From Sect. 4.5 we know when we rest at some point in an inertial state, our clock is running *faster* than all clocks that first rest with us, then move away a little bit and return to us, resting again with us, after some time. Now the equivalence principle comes into play. Being in the rocket in Fig. 5.4, we do not feel any force pulling us, while free-falling along the solid curved line from A to B. Hence we can insist that we *rest*. Therefore any clock that after resting with us has moved a little bit away at point A and returned at point B, and is then again resting with us, will run more slowly than our clock. Likewise, being in the rocket *we can* move along the dashed line, but we must then switch on the rocket motor and *accelerate*. We must leave our present inertial state. Before leaving, we place a clock next to our rocket which continues to free-fall from A to B, along the solid curved line. If we meet again and rest relative to this clock at B, then our clock is running *more slowly* compared to the clock that traveled along the bended path and that is free-falling in an inertial state.

Next we need to know what a "nearby" path in space-time means by analogy with Fig. 5.7. If we move on the original path and our friend moves along a nearby path, then he must at "short-enough" intervals *rest* for a moment beside us. We call such a path a **space-time nearby path**.

For the "length" of a path in space-time, we can use the *proper time* needed to pass through it. The *straightest* path between *A* and *B* is the path taking *the most* proper time of all *space-time nearby* paths and having the same starting and arriving speed.

On the analogy of a straightest but not shortest path on a surface in Sect. 5.2.1, can it happen that while traveling on such a straightest path our proper time is running *more slowly* than it does for our friend who left us some time ago, and just joined us again? This is possible if our friend does not follow a space-time nearby path, as we will see in Sect. 6.2.

Such space and time are connected to space-time. Just drawing some line between two points *A* and *B* in *space* does not mean anything. In addition, we must fix the starting speed and, at least in a thought experiment, we *travel* from *A* to *B* to find out what the *straightest* path is. It will be a *path* on which we can travel force-free, that is **free-falling**. The time needed depends also on the starting speed in bended space-time, and we use it as the "length" of the path in space-*time*.

All matter is free-falling in the same way if, *at the same time*, it is light enough so that its gravity does not practically influence other masses, and small enough so that its movement traces out a line. We call such a mass a **test mass**, such as for example a small clock. Under gravity, the test masses will move relative to each other and show us how a piece of space *deforms* as time proceeds. So we can see how space-time is bending. We explain in the next section how this works.

Let us sum up:

> Gravity bends space-time. The **straightest path** between two points is when a test mass is **free-falling** between them, that is when the **proper time** for a test mass traveling between the points at a given initial speed is passing **fastest, compared with space-time nearby paths**. This path is a **geodesic in space-time**.

5.3 Measuring the Bending of Space-Time

We said in Sect. 5.1 that to detect gravity we have to look outside the elevator. All the same, we can use a spaceship being large enough and free-falling to Earth without air inside, as we sketched in the left-hand picture of Fig. 5.8.

On the left and the right hand of the astronaut there are two balls falling together, parallel to him. However, gravity acts in the direction of the center of Earth. Hence the two balls begin to head towards the center of Earth and are approaching each other. In the picture on the right, we see that for the astronaut the two balls will begin to move *towards* each other.

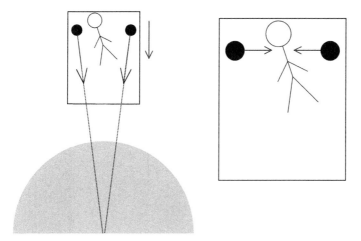

Fig. 5.8 Balls inside a spaceship, large enough and free-falling, begin to move *towards* each other

Fig. 5.9 Moving parallel
along two straightest lines on
a ball, two observers can
move towards each other

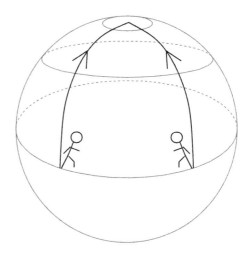

Again, we can draw an analogy with the straight-ahead motion on the bending
surface of the Earth. In Fig. 5.9, two observers starting at the equator and moving
straight-ahead northwards will meet at the North Pole. So they move towards each
other.

However, there is more to this analogy: in Fig. 5.10, two observers move straight-
ahead towards the equator. Hence they move away from each other.

An analogous thing happens in the spaceship. In the left-hand picture of Fig. 5.11,
we placed the balls in the free-falling spaceship on top of each other. The upper ball
is farther away from the center of Earth than the lower ball so that gravity acts more
strongly on the lower ball.

Fig. 5.10 Moving parallel
along two straightest lines on
a ball, two observers can also
move away from each other

Fig. 5.11 Balls inside a
large, free-falling spaceship
on *top* of each other begin to
move *away* from each other

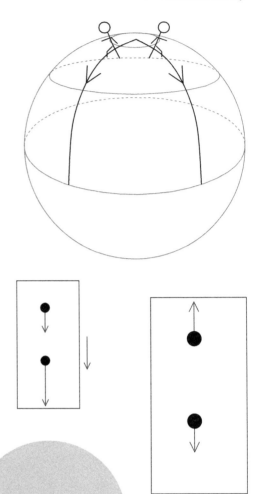

Inside the spaceship, in the right-hand picture of Fig. 5.11, we observe therefore
that the upper and lower ball will begin to *move away* from each other!

How does the whole picture fit together? Let us place even more balls as test
masses inside the spaceship, resting relative to each other, so that they mark the
corners of an *imagined* shoe box. Again we suppose by way of simplicity that there
is no air inside the spaceship. We sketched one side of the box in the left-hand picture
of Fig. 5.12 in light gray. In the center of the picture we see both effects working at
the same time. The left and right balls move towards each other while the upper and
lower balls move away from each other. In other words, the imagined box will gain
height but lose width, and in the third dimension depth, as we see in the picture on
the right.

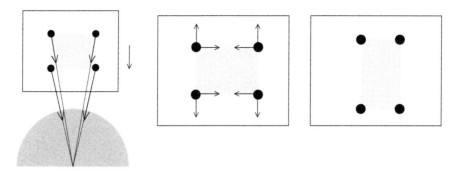

Fig. 5.12 Inside a large, free-falling spaceship the form of a small piece of space begins to change but not its volume

Now if you measure this carefully, you will find that the *volume* of the box does not change! Can we understand this intuitively? If the imagined box contains mass, then the box would *begin to shrink* because gravity tends to clump mass. However, we marked a box with *test masses* that do *not* disturb the neighborhood. Hence the box *contains no mass*. The gravity acting from *outside* masses like Earth will *deform* the box, but *not shrink* it.

This is the essence of what gravity does. We will show how mass *exactly creates* gravity in Chap. 7. However, at first we have to understand in more detail how mass *reacts* to gravity.

Chapter 6
Equivalence Principle in Action

6.1 Time and Gravity

How does gravity influence time? In order to answer this question, we compare the pace of two identical clocks *resting relative* to each other, as sketched in Fig. 6.1. The right clock is resting on the surface of the gray planet. The left clock is resting far enough away from the planet so that we can neglect the planet's gravity. Therefore the left clock is *nearly* in an inertial state.[1]

We arrange that at some time both clocks show say, two o'clock, as you see in Fig. 6.1. We ask: does the right clock on the planet advance at the same pace as the left clock that is far away from the planet does?

In order to compare the pace of the two clocks, we move the left clock to the right clock *without* changing its inertial state. This is possible because the equivalence principle tells us that a free-falling clock *is* in an inertial state. Therefore we choose to free-fall together with the left clock towards the planet starting at, say two o'clock. Then for us, the planet with its clock on it is moving *towards* us. The horizontal arrows show the speed. In Fig. 6.1, the planet with the right clock resting on it just starts to move *relative* to us. In the next Fig. 6.2, the right clock has nearly reached us. When we pass near the right clock, we can compare the two clocks. Our clock was always in the same inertial state, so it is proceeding at the *same* pace as it did far away from the planet. However, we see that the clock on the planet is moving relative to us at a certain speed when we pass it. Hence we conclude that the clock on the planet *is running more slowly* against our clock.

How much is the right clock running slowly against the left clock? This is determined by the γ factor stemming from the *speed* with which the left clock is passing the right clock. This speed depends on the radius and the mass of the planet. We use this thought experiment to *calculate* the slowing rate in Sect. 8.5 when we solve the Einstein equation of gravity.

---

[1] For the meaning of "far enough" or "nearly", see the last section of the preface.

© Springer International Publishing Switzerland 2015
K. Fischer, *Relativity for Everyone*, Undergraduate Lecture Notes in Physics, DOI 10.1007/978-3-319-17891-2_6

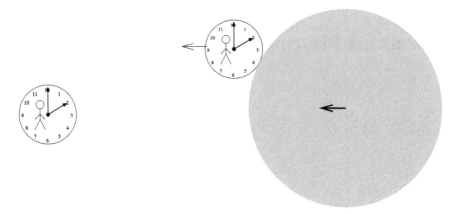

Fig. 6.1 The *left* clock begins to free-fall from far away towards the planet

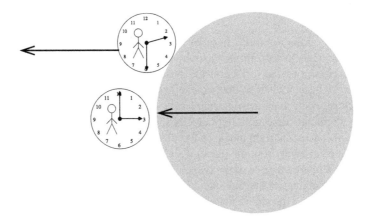

Fig. 6.2 The *left*, free-falling clock passes the *right* clock which rests on the planet

Let us sum up:

Gravity of a mass slows down nearby clocks. If we let another clock start from a place resting far away from the large mass free-falling towards it and passing a clock resting near that large mass, then the γ factor of this speed is the rate at which the clock resting near the large mass is running slowly.

Some popular texts suggest that it is the amount of *acceleration* we feel on the planet which is slowing down the clocks. They invoke the equivalence principle and compare the clock on the planet with a clock in a starting rocket. This is incorrect. Compare with the merry-go-round of Sect. 4.5. We can build two merry-go-rounds, one with a smaller diameter and one with a larger diameter. We let them both rotate

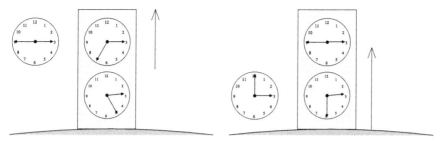

Fig. 6.3 In a high-rise building clocks on lower floors run more slowly than clocks on higher floors

such that the rim moves at the *same* speed. Then the *acceleration* on the rim of the small merry-go-round is *larger* than on the rim of the larger merry-go-round, but the time delay of clocks at the respective rim is *the same*. Likewise we can find two planets, one with roughly double the mass and diameter of the other, such that they have a *different* acceleration due to gravity, but the *same* time delay of clocks due to the fact that the left clock will pass the right clock resting on the planet at the same speed. We will learn the exact condition in Chap. 8.

In order to see the effect on Earth, we just use two very precise clocks in a high-rise building. We adjust the clocks and put one clock on the ground floor of the building and one on the upper floor. Then we free-fall again from far away and pass at the high-rise building, as in Fig. 6.3. We pass the upper clock at some speed and the lower clock at some larger speed. Hence we see that the upper clock is proceeding at a faster pace than the lower clock. In other words: we can observe directly that time on the upper floor of the high-rise building is proceeding faster than on the ground floor (without having to free-fall!). In fact, recent advance in clock technology made it possible to observe this effect already for a height difference of about 33 centimeters![2]

6.2 Proper Time in Bended Space-Time: Twin Paradox 3

In Sect. 5.2.2 we learned that, compared with space-time *nearby* paths, the proper time of a free-falling test mass is passing fastest. However, is this also necessarily true if two twins who were first together move along *completely different* paths in bended space-time? For this please have a look at Fig. 6.4. At first, the two twins free-fall around a planet as you can see in the left-hand picture of Fig. 6.4. Then the white-headed twin accelerates and stops at the platform, as in the right-hand picture.

The white-headed twin sees again and again the dark-headed twin passing by him, encircling the planet and free-falling. Finally, when the dark-headed twin passes the platform again, the white-headed twin rejoins the dark-headed twin. While the

---

[2]For example, in 2010 this has been checked by C.W. Chou, D.B. Hume, T. Rosenband, and D.J. Wineland from the National Institute of Standards and Technology, USA. See for example: http://www.sciencedaily.com/releases/2010/09/100923142436.htm.

Fig. 6.4 The dark-headed twin continues to free-fall in a circle around a planet, while the white-headed twin stops at the platform

Fig. 6.5 We are carrying a clock and at the same time we are passing vertically the white-headed twin standing on the platform and the free-falling dark-headed twin. The *arrows* show the speed, as seen by the white-headed twin

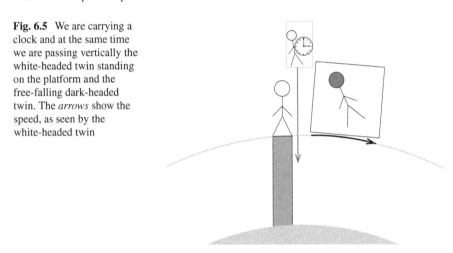

dark-headed twin continued to free-fall, the white-headed twin was constantly accelerating: he clearly accelerated in order to depart from and to rejoin the dark-headed twin. What is more, while he was standing on the platform the white-headed twin constantly felt the gravity of the planet, that is he constantly accelerated also there.

Let us compare the proper time of the two twins. We carry a clock and start free-falling vertically from a resting place far away from the planet. We fall such that we pass both the dark-headed and the white-headed twin at the same time, as you see in Fig. 6.5. We see the white-headed twin moving upwards towards us at some speed. However, the dark-headed twin moves in addition to the vertical speed also at some horizontal speed, that is *faster* than the upwards moving white-headed twin relative to us. Hence the proper time of the dark-headed twin passes *more slowly* than the proper time of the white-headed twin standing on the platform.

During the stopping and starting from the platform, the proper time of the white-headed twin will change somehow. Because the white-headed twin can wait on the platform as long as he wishes, he can make sure that when he joins the dark-headed twin again that his proper time proceeded *more* than the dark-headed twin's proper time.

This is again an example of the **twin paradox** or **clock paradox**. However, in contrast to what happened in Sects. 4.1 and 4.4, the time of the twin remaining in the inertial state proceeds *more slowly* than the time of the twin who changes his inertial state.

In other words: although the dark-headed twin moved along the *straightest* path in space-time between leaving and joining the white-headed twin, his proper time of all paths did not pass fastest. This is analogous to the straightest westward path along the equator in Fig. 5.6. Similar to that figure, the dark-headed twin does not move on a **space-time nearby path** to the white-headed twin, but "westwards".

6.3 Moving Straightly in Bended Space-Time

In Sect. 6.1, the clock free-fell along a straight line. However, in bended space-time this is a curved line, analogous to Fig. 5.4. Let us see why. In the left-hand picture of Fig. 6.6, we sketched a very light planet as a gray disk. This planet creates nearly no gravity. For a better view, we drew the planet smaller than the clock. The two dimensions in this picture are not width and height but *time* and distance in space-time. We see that the clock moves towards the light planet along a *straight* line in space-time at an even pace. When one third of the time span has elapsed until impact time, the clock traveled one-third of the distance to the planet. When two-thirds of the time has elapsed until impact time, the clock traveled two-thirds of the distance and so on.

In the picture on the right-hand side, we have a situation similar to the one in Fig. 6.1. Now the planet has mass and we sketched this as a dark black disk. We let the clock start at the same initial speed as in the left-hand picture. The clock will **accelerate relative** to the planet while approaching it. It now follows the curved solid line instead of the straight dashed line and reaches the heavy planet in *less* time than the very light planet before. Although the clock free-falls along a straight line in *space* towards the center of the planet, the clock moves along a bended path in *space-time* near a gravitating mass.

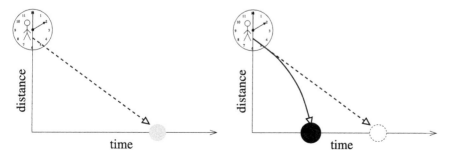

Fig. 6.6   Bended path in space-time

6.4 Length Under the Gravity of a Perfect Ball

In order to see how lengths change under gravity, let us **model a planet or star**. For
physical phenomena the physicists try to construct the simplest of possible models
that still contain the *essence* of the phenomenon. Let us do the same. We know that
planets or stars are nearly balls. Their mass may vary with depth but in all directions
there is nearly the same amount of mass. This is sketched in Fig. 6.7. This is a good
assumption for real planets and stars. We call such a body a **perfect ball**.

Let us imagine the surface of a ball in empty space which has the same center
as the perfect ball. We sketched it as dashed line in Fig. 6.7. Next, several observers
who were resting relative to the ball at the same large distance, free-fall towards the
perfect ball. They measure time and length in this inertial state. Because the mass of
the perfect ball is the same in all directions, gravity acts the same in all directions,
and the observers free-fall at the same pace. Therefore they pass at the same speed
through the dashed surface of the imagined ball.

Let us free-fall with one of the observers. Then this dashed ball moves relative
towards us. A friend resting at the imagined dashed ball surface has placed rods
along its surface. We sketched this in Fig. 6.8. The lengths *along* the surface of the
dashed ball are at right angles to our speed relative to the perfect ball. Hence when
we pass such a rod, the length of the rod does not change by this speed, as we saw in
Sect. 2.4.2. All observers agree on that. Lengths on the surface of *any* ball with the
same center as the perfect ball do not change. In particular, the length of *any* circle
with the same center as the perfect ball is the same as it would be without gravity.

Fig. 6.7 In a perfect ball,
mass density depends maybe
on depth but not on direction.
Darker colors mean layers of
higher mass density

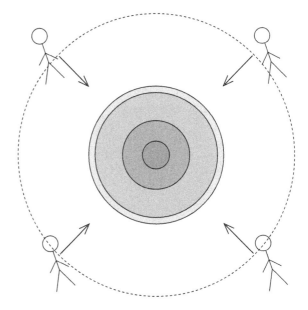

Fig. 6.8 Around a perfect
ball, rods along the diameter
shrink, but rods at right
angles do not, according to
an observer free-falling from
a resting place far enough
outside

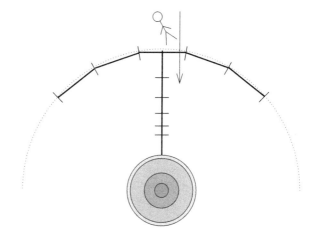

Our friend resting at the dashed imagined ball also placed identical rods along the
diameter of the ball. If we pass such a rod, it is in the direction of our speed which
grows as we approach the perfect ball. Hence relative to us, this rod shrinks by the
γ factor of our passing speed as we saw in Sect. 2.4.1. We gained this speed while
free-falling from a far-away resting point towards the perfect ball. However, we are
still in the same inertial state as when we started resting far away from the planet.
Hence relative to an observer who is resting sufficiently faraway from the planet, a
rod resting with the planet along the diameter shrinks more and more the nearer it is
to the center.

What does our friend resting near the perfect ball observe? He will need *more*
rods to exhaust the diameter than he would need without the gravity of the perfect
ball. Because of gravity, the ratio of the length of any circle with the same center as
the perfect ball and its diameter will be *less* than π. Hence, near a perfect ball, space
itself is bending!

It will turn out that there is a more practical way to keep track of the bending
space. In Fig. 6.9 the observer rests between two circles with the same center as the
perfect ball. He counts the number of identical rods needed to trace the outer circle
and subtracts the number of rods needed to trace the inner circle. If space does not
bend, the number of rods on the outer circle is the number of rods needed to trace its
radius $\times 2\pi$. Likewise, the number of rods on the inner circle is the number of rods
on its radius $\times 2\pi$. Hence the number of rods fitting into the solid line between the
two circles, in front of the observer in the figure, is the difference of the number of
rods on the large and the small circles, divided by 2π. The observer wants this ratio
to remain *the same* even when a perfect ball in its center is bending space. We know
that the rods on the circles will not change their length, even with the perfect ball
in its center now gravitating. However, the rod in front of the observer in the figure
drawn as a solid line will shrink by the factor γ, that is corresponding to the speed of
the bypassing clock free-falling from a resting place sufficiently far away from the
perfect ball.

Fig. 6.9 Our friend standing between the two nearly equally large *dashed circles* needs *more* rods to fill the *solid line* than when there is no gravity, by the factor $1/\gamma$ of the speed of the clock free-falling from a point resting far away from the perfect ball

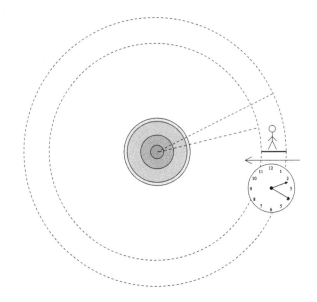

Hence the observer decides to *stretch* the rods in front of him by the factor $1/\gamma$ so that now the number of *stretched* rods is again the difference of the number of rods on the large and the smaller circle, divided by 2π. The bending of space now becomes apparent because the distance from the center of the perfect ball grows *faster* than the number of rods along the diameter. In other words: The distance to the center is *larger* than the such measured radius. As for us free-falling from a resting place sufficiently far away from the perfect ball, the rod shrinks by the same γ factor so that for us, the rod has now the *same* length as the *original* rods far away outside. This is the reason why this way of measuring is practical.

6.5 Gravity Around a Perfect Ball

Let us sum up what we learned about **gravity around a perfect ball** of mass. Space-time bends in the following way:

1. At a place resting near the perfect ball, time will run more slowly by a factor γ, relative to a place resting sufficiently far away from the perfect ball.
2. This γ factor belongs to the speed that the clock reaches here, after free-falling vertically from a place at rest sufficiently far away from the perfect ball.

3. This factor γ depends only on the radius. Far outside, γ is one. The smaller the radius is outside the perfect ball, the smaller γ.
4. Lengths at right angles to the diameter will not change. Geometry on the surface of a ball with the same center as the perfect ball does not change.
5. The ratio of the perimeter and the radius of any circle with the same center as the perfect ball will be 2π, and the surface area of a ball with the same center as the perfect ball will be 4π times the *squared* radius as in basic geometry.
6. For an observer resting near the perfect ball, the distance between two nearby points on the same diameter will be *larger* than what the measuring tape shows, by the factor $1/\gamma$. In other words: the radius measured in a number of rods is *smaller* than the distance to the center of the perfect ball.

   However, for an observer free-falling from a resting place sufficiently far away from the perfect ball, the measuring tape shows the correct distance.

This manner in which rods and time change locally is called a **metric**. It allows us to *measure* how space-time is bending. We see that the equivalence principle gives us nearly the complete picture of how gravity is bending space-time around a perfect ball. The only information we are still missing is how exactly the shrinking factor γ is depending on the radius! We will calculate this factor in Chap. 8, when we solve the **Einstein equation of gravity** *exactly* for the perfect ball. This bending of space-time is the **Schwarzschild solution** of the Einstein equation of gravity or **Schwarzschild metric**. It is named after **Karl Schwarzschild**, a German astronomer and physicist. It is the most important exact solution of the Einstein equation of gravity because planets and stars are nearly perfect balls after all.

6.6 Mass Under Gravity

We saw in Sect. 2.7 that if clocks slow down, then inertial mass increases. Hence a gravitating mass increases the **inertial mass of a test mass nearby**. This is remarkable. There is so much mass in the universe acting on a test mass with its gravity so that we may speculate:

What if *all* inertial mass of bodies comes from the gravity of the other masses in the universe?

This is not a strictly physical law but only an idea called the **Mach principle**. In fact, even before the theory of relativity was created, Ernst Mach speculated about this possibility. It is interesting because it would allow us to state one *reason* why matter has mass. However, it never led to an exact theory. The equivalence principle shows that at least some part of the inertial mass of a body may come from nearby gravitating masses.

6.7 Light Under Gravity

Suppose that a light beam passes through a transparent free-falling box, as in the left-hand picture of Fig. 6.10. The box and the observer in the box are in an inertial state. Therefore the observer in the box sees the light passing straight away through the box. Then the equivalence principle tells us that in the picture on the right, the observer on the planet will see the box accelerating towards him together with the light beam. The light beam *bends*, that is it **accelerates** towards the planet. This effect is strongest if the light passes near the surface of the planet or the star because there gravity is strongest.

The sun has enough mass so that we can see this effect: we take a photograph of distant stars behind the sun so that their light bends near the surface of the sun. However, the sun is much too bright so that we have to wait for the moon interrupting the light from the sun, at a solar eclipse, as in Fig. 6.11. Here the small dark moon stands between us and the large gray sun in the center. The light-gray drawn stars really stand more closely together than they look on the photograph, such as "white" stars.

To verify this, we wait half a year. Then, at night the sun is on the left behind us as you see in Fig. 6.12. We take the same photograph again. This time we see the stars at the place of the "gray" stars. We see that the stars are really nearer together than they seemed to be when the sun was in-between:

Light bends near mass because space-time bends.

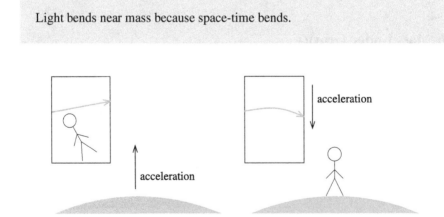

Fig. 6.10 Light, depicted as *light-gray arrow*, bending under gravity

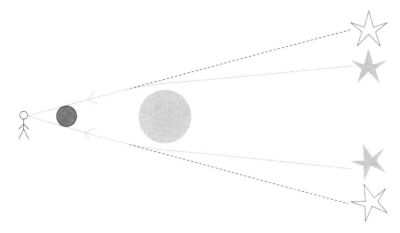

Fig. 6.11 Light bending along the sun

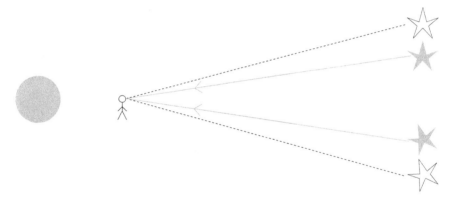

Fig. 6.12 We see the original position of the stars half a year later

In this setup, the stars are very far away from the bending body, and the observer is relatively nearby.

Another possible setup is when the bending body, a yellow galaxy for example, is far away and a blue star is behind the galaxy as you see in Fig. 6.13. The galaxy acts as **a gravity lens**. Then we can see the light from the blue star as a *ring*. This is the so-called **Einstein ring**. In Fig. 6.14, you see a photograph of such a ring.[3]

In Sect. 9.2 we calculate the angle at which light bends while passing a star by using the Schwarzschild exact solution.

---

[3]Credit: NASA, ESA, A. Bolton (Harvard-Smithsonian Center for Astrophysics) and the Sloan Lens Advanced Camera for Surveys Team.

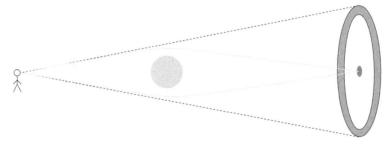

Fig. 6.13 A galaxy acts as a lens for the light from a distant star

Fig. 6.14 Einstein ring with
the number SDSS
J162746.44-005357.5,
photographed by the Hubble
space telescope. The
resolution you see here, is
the resolution of the camera

6.8 Black Holes: A First Look

The more mass a star contains, the more gravity acts on its neighborhood. Light
bends more, time runs more slowly, inertial masses get more inertia. Hence, maybe
from a certain mass on, light rays bend so much that even light sent upwards from
its center is falling back: the star emits no light. Such a body is indeed possible: it is
called a **black hole**. Around it there is a ball-shaped imaginary surface, the **horizon**.
Any body can travel from the outside through this horizon, but as for a trapdoor,
there is no way out of a black hole.

If some spaceship approaches the horizon from the outside, then for a far-away
observer time in the spaceship is running slower and slower. When it reaches the
horizon, time is running *infinitely slowly* as you can see in Fig. 6.15. We never *see*
the spaceship traveling *through* the horizon. However, the spaceship is free-falling,

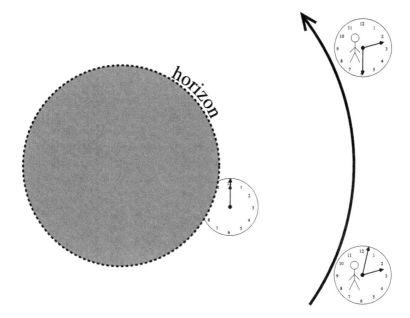

Fig. 6.15 Clocks run more slowly near large masses and freeze at the horizon of a black hole, relative to outside observers

so it does not notice anything special in its neighborhood: it will travel through the horizon into the interior of the black hole. This is because the equivalence principle tells us that free-falling test masses act as if they are floating in empty space.

There is no contradiction. An astronaut in the spaceship *cannot tell* the outside world that he has passed inside the horizon because *not even light* can pass through the horizon from the interior.

We will calculate the necessary mass of a black hole on the basis of the Schwarzschild exact solution in Sect. 9.1.

6.9 Equivalence Principle: Summary

If we look at gravity just as a force that pulls masses together, then why on earth should the inertia be the same as the weight of the mass? There is no *reason* for it. Einstein saw that this equality of inertial and gravitational mass is the key to understanding gravity more deeply. All small enough mass *reacts* in the same way to gravity. Because the inertia resists acceleration and the heaviness tends to accelerate and both are of exactly the same size, test masses in gravity will not accelerate at all, but move in an inertial state, force-free, that is, free-falling.

Moving force-free, that is free-falling does not depend on the material. This explains why inertia and heaviness are always of the same magnitude.

However, we know that test masses move under gravity along *bended* paths at varying speed, so gravity must *bend* space and time. Gravity does not bend space alone because the free-falling test masses do not follow the shortest path in space but the path that takes the *longest proper time*, as compared with nearby paths in space-time. This path is called a **geodesic in space-time**. We can produce a geodesic by fixing the direction *and* the starting speed of the test mass and letting it free-fall. Different starting speeds give different geodesics. So, again it is not only space that bends. Based on such **free-falling** test masses, we can use our knowledge about the theory of special relativity to understand how mass **is reacting to gravity**.

Why does matter create gravity? Nobody knows! We only know *how* matter creates gravity. It is simply consistent with the equivalence principle in a *simplest possible* way. We will see this in the next chapter.

Chapter 7
How Mass Creates Gravity

From experience we know and we learned in the preceding chapter that mass creates gravity. We saw that gravity bends space-time, so mass *itself* should bend space-time. What is the simplest possible way that mass *can* bend space-time?

7.1 Gravity in a Lonely Cloud

We simplify the situation as much as possible. We drive with our spaceship into some empty region of space so that no large mass is nearby, and we are in an inertial state. Then we carefully place outside the spaceship a small cloud of dust such that the dust particles are resting near each other. In Fig. 7.1 we sketched the dust particles as black balls. Then we gently move away and sit motionless near the cloud. The cloud as well as we are in an inertial state.

From experience we expect that gravity tends to move mass together. In fact, as soon as we leave the dust particles alone, the cloud begins to **shrink**. What is the simplest quantity to describe the rate of shrinking? This is how much the **volume** of the cloud decreases per *time*, that is, at which speed the volume decreases. What is the simplest quantity to describe the *initial rate* of shrinking? That is how much the shrinking speed starts to change from zero which means the *acceleration* of the volume decrease and the change of volume per time, *per time*. The start of the shrinking rate should depend on how much mass is in the cloud. In the simplest case, the rate should grow *in proportion* to the mass in the cloud. Twice as much mass in the same volume should produce twice as large shrinking rate.

Indeed, this is what nature has chosen as the law of gravity!

Let us give a numerical example of how a volume may begin to shrink. Suppose that we prepare a dust cloud of the size $10 \times 10 \times 10$ meters, that is a volume of $1000 \, \text{m}^3$. When this cloud begins to shrink, we measure its volume each second. In Fig. 7.2, we see in the column of boxes on the left how much volume is left after $0, 1, 2, 3, 4$ seconds. The rate at which the volume shrinks per time, that is from second to second, stands in the middle column of boxes. Finally, we find in the right

© Springer International Publishing Switzerland 2015
K. Fischer, *Relativity for Everyone*, Undergraduate Lecture Notes
in Physics, DOI 10.1007/978-3-319-17891-2_7

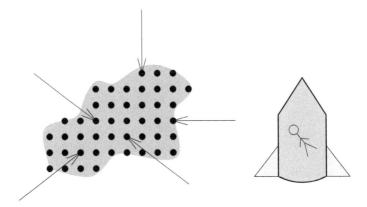

Fig. 7.1 A dust cloud begins to shrink under its own gravity

Fig. 7.2 Example of how a volume may begin to shrink

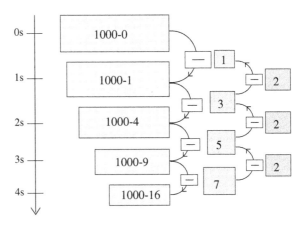

column the rate at which this change of the volume *itself* is changing per second, that is the change per second. We see that the volume *begins* to shrink at a rate of $2\,m^3$, per second, that is $2\,\dfrac{m^3}{s^2}$.

However, we wanted to know how mass bends space-time, not just how mass clumps. Here the equivalence principle comes again.

We know that test masses, especially small dust particles, will *react* to gravity in the *same* way, no matter what their matter consists of. Therefore we can think of the dust particles as just *probing* the space-time around them. While the cloud is beginning to shrink, the test masses *accelerate* towards each other. However, remember that nothing pulls the test masses together. The test masses accelerate towards each other because inside the cloud, space *itself* is beginning to shrink in time. Hence when the cloud begins to shrink, the **small volume of space** itself begins to shrink, as **time** proceeds a little bit. This is how mass is bending a small piece of space-time.

We have found the Einstein law of gravity!

7.2 Einstein Equation of Gravity

The **Einstein law of gravity** or under another name the **Einstein equation of gravity**, is:

> The rate at which a small enough, resting cloud of matter begins to shrink is in proportion to the *mass* in that cloud. The constant of proportion is 4π times the **gravity constant**.

Why not just "gravity constant" but 4π times it? This has purely historical reasons, nothing else. The value of the gravity constant is about 6.67×10^{-11} in appropriate units. You find it for reference in the Table A.1.

The **mass density**, that is the mass per volume, is nearly constant if we look at a small enough volume. It often is more practical to ask for the shrinking rate *per volume*, that is the **relative shrinking rate**. The **Einstein equation of gravity in terms of the mass density**, that is in terms of the mass per volume, reads

> The *relative* rate at which a small enough, resting cloud of matter begins to shrink grows in proportion to the *mass density* in that cloud. The constant of proportion is 4π times the gravity constant.

However, mass is energy, divided by the square of the speed of light. In terms of the energy density, the **Einstein equation of gravity for energy** reads

> The *relative* rate at which a small enough, resting cloud of matter begins to shrink grows in proportion to the *energy density* in that cloud. The constant of proportion is 4π times the gravity constant, divided by the square of the speed of light.

7.3 Enter Pressure

We assumed that we can always place our test masses inside the cloud, such that, at least at the beginning, they rest with each other. What happens if our cloud does also contain **pure energy**, that is light? Pure energy moves always at the speed of light, so we cannot place it like our masses. This light behaves much like a gas. Suppose we put our masses inside a gas. The simplest case is when at least near the cloud the gas looks everywhere the same. Therefore the gas particles, or light in the case of pure energy, are constantly entering and leaving the cloud which we marked with test masses. The number of gas particles which leave the cloud per second is the same as the number of particles which enter the cloud during this second. Still, the particles of the gas always bump against each other so that they never rest relative to the test masses in the cloud. You can see this in Fig. 7.3. Because just as many gas particles enter the cloud per second as they are leaving the cloud, we can imagine that for every gas particle that is about to leave the cloud, a gas particle from the outside bounces against it to keep the gas inside the cloud together. In other words: the outside gas **presses** from any of the three directions of space. But pressure is energy density, as we saw in Sect. 1.12. This means that we have to add the sum of the three pressures in the three directions of space to the energy density in the cloud.

As a result, we get the **complete Einstein equation of gravity**:

The *relative* rate at which a small enough, resting cloud of matter begins to shrink, grows in proportion to the *energy density* plus the *pressures* in each of the three directions in that cloud. The constant of proportion is 4π times the gravity constant, divided by the square of the speed of light.

However, inside the cloud, the gas particles are not only moving around but also bumping against our test masses. If pressure is constant in all directions, the test mass will not move on the average. However, under a microscope we see the test mass vibrating under the bombarding small gas particles. Hence we cannot place our test masses *perfectly* at rest. The very concept of **pressure** does only make sense if we do not look *too* closely, that is if our cloud of masses is not *too* small.

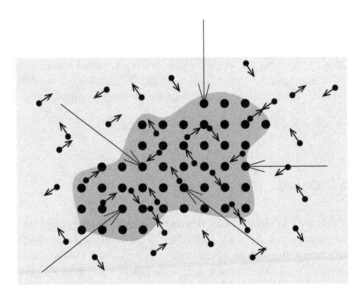

Fig. 7.3 A dust cloud inside a gas. We sketched the gas particles as *small black disks* and we show their momentary speed by *small arrows*. The dust cloud contains also the energy coming from the disordered moving particles in the gas. This energy also creates gravity

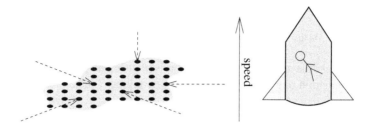

Fig. 7.4 A moving dust cloud begins to shrink

Einstein himself was aware of this[1]:

> We know that matter is built up of electrically charged particles but we do
> not know the laws which govern the constitution of these particles. In treating
> mechanical problems, we are therefore obliged to make use of an inexact
> description of matter which corresponds to that of classical mechanics. The
> density [. . .] of a material substance and the hydrodynamical pressures are
> the fundamental concepts upon which such a description is based.

With the exception of Sect. 9.9, we deal in the following sections with matter
whose pressure is so small that we can ignore it so that we can still use the model of
the dust cloud of the Sect. 7.1.

7.4 Enter Speed

The law of how mass creates gravity should fit in with the theory of special relativity.
That is: if we are in an inertial state and the cloud is passing us free-falling at some
speed, then the Einstein equation of gravity should not change. However, we know
that all kinds of things change. First of all, the cloud has more mass by a factor $1/\gamma$.

Therefore it creates more gravity! However, also the reaction of the volume
changes. The length of the cloud in the direction of the speed is now smaller by
this γ factor, while its size vertical to the speed does not change, as we can see by
comparing the Figs. 7.1 and 7.4. Hence the *volume* of the moving cloud is smaller
by this factor γ. What is more, the volume begins to shrink in *less* time. We know
from Sect. 2.1 that our time runs faster than the proper time of the moving cloud by
the inverse factor $1/\gamma$. Hence the cloud shrinks faster by this amount and it *begins*
to shrink even at the faster rate of $(1/\gamma) \times (1/\gamma)$. In total the volume of the cloud
begins to shrink faster by the factor

---

[1] A. Einstein. *The meaning of relativity*. Princeton University press, 1956.

$$\underbrace{\gamma}_{\text{Volume shrinks}} \quad \times \quad \underbrace{(1/\gamma) \times (1/\gamma)}_{\text{shrinking begins faster}} \quad = 1/\gamma$$

which is precisely the amount by which the mass in the cloud became larger. Hence the Einstein equation of gravity holds also for free-falling clouds.

7.5 Enter Outside Masses

In Sect. 7.2 we described how mass creates gravity for a carefully prepared small cloud of dust, far away from other large masses. In reality there are stars, planets and the like *outside* the cloud. Let us consider them. Because our cloud is small enough, it moves like a test mass under the gravity of outside masses, that is, it free-falls as you can see in Fig. 7.5. Based on the equivalence principle, the cloud is free-falling and reacts *as if* there would be no large mass nearby. The shrinking rate does only depend on how much mass is *inside* the dust cloud. The law of gravity does *not* change at all!

The outside masses can only change the *form* of the small free-falling cloud of mutually resting test masses, as we already saw in Fig. 5.12.

7.6 Local and Global Space-Time

The Einstein law of gravity is formulated for a *small enough* piece of space (our "cloud") and a *small enough* period of time in which the cloud "begins" to shrink, that is altogether in *a small enough* piece of space-time, thus a *local* piece of

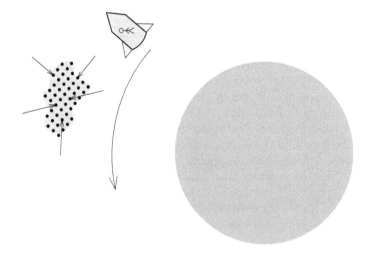

Fig. 7.5 A shrinking small cloud acts like a test mass and free-falls under gravity of a nearby mass

space-time. How do we get the *global* picture of what happens under gravity? Take for example the sun. We know that the sun will bend space-time and that far away from it, space-time will be *flat*. We insert here and there small clouds of test masses to probe space-time. Then we *patch* these local pictures to a smooth map of space-time. It is a little like building a globe of patches showing parts of the surface of Earth. Only if we have *connected* them smoothly do we see that Earth indeed has the form of a globe. We get the *global* picture of how the globe is bending.

However, in space-time, things are more involved than for a surface. If the volume begins to shrink near a mass, then space-time must bend to connect the nearby space where there is no mass. What is more, not only does the volume of a cloud shrink in time, but time *itself* depends on the relative speed of the free-falling masses in a cloud. So we have to follow how *space and time* evolve.

7.7 Bended Space-Time and Tensors

For a bended surface in an ordinary three-dimensional space, it is enough to know one number per surface point, the **Gauss curvature** to understand how a surface bends. This curvature basically measures how much the ratio of the perimeter and the diameter of a small circle around a point differ from π, as in Fig. 7.6. In other words, it measures to what extent the Euclidean geometry is wrong. However, space-time has more directions in which it can bend so that one number is not enough. Careful thinking shows that for each point in space-time we need *twenty* numbers to render the bending rate.

It was Riemann who generalized the theory of bended surfaces of Gauss to three and more dimensional spaces. The twenty numbers are named after him, the **Riemann curvature tensor**. For instance, for our cloud of dust particles of Sect. 7.1

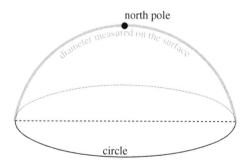

Fig. 7.6 Measuring how much the surface of the Earth is bending near the North Pole. Draw a circle around the North Pole, on the surface of Earth. The perimeter of the circle is π times the *black dashed* diameter which runs inside the Earth. However, the diameter measured *on the bended surface* of Earth is the gray arc spanning from *left* to the *right*, and this is *longer* than the *dashed straight line*. Hence on the surface of the Earth, the ratio of the perimeter and the diameter of a circle is *less* than π, similar as in Fig. 6.8 for a circle around a **perfect ball**

which at first are resting near each other, the Riemann curvature tensor describes how the *form* and the *volume* of the cloud is beginning to change as time proceeds.

The Einstein equation of gravity describes only how the *volume* of the cloud is beginning to change but not how its *form* does. In order to express the Einstein equation of gravity in terms of tensors, we need therefore a simplified version of the Riemann curvature tensor, namely the **Ricci tensor**.

The mathematical toolbox for such calculations is **tensor analysis**. "Tensor" is a word of Latin origin and means "tension". In fact, engineers use tensors to describe how materials of bridges and the like bend under *tension*. Materials bend *inside* space-time which we can easily show, as in Fig. 7.7. However, it is much harder to imagine how space-time *itself* bends. We will show in Sect. 9.7 that there is a fundamental difference between a bending material and a bending space-time.

If you like to see the mathematical expression for the Einstein equation of gravity, in terms of the Ricci tensor, please have a look at Appendix A.5.

7.8 How to Solve the Einstein Equation of Gravity

Not only obtaining the global picture is difficult but focusing on one small cloud is difficult as well. What happens *after* that cloud has begun to shrink? The test masses inside have just begun to move against each other, free-falling. To follow them, we must look into even smaller parts of the cloud, adjust our speed and time, look for the shrinking rate and so on. That is because the masses in the cloud *act* on space-time at the same time by bending it, and *react* on space-time by free-falling. We cannot easily separate action and reaction of mass. In other words, not every pattern of masses about which we think can be realized in space-time. That makes gravity unique among the interactions in nature. That's just what makes solving the Einstein equation of gravity so difficult.

In fact, we know only a few exact solutions to the Einstein equation of gravity. In order to solve it *exactly*, we must place mass in a balanced way so that we can handle the self-interaction.

1. For the **perfect ball** of mass, as in Sect. 6.5, the same amount of mass is sitting in every direction from its center. Hence mass balances out. Gravity depends only on the distance to the center of the ball, not the direction. All that remains to

Fig. 7.7 A material bends

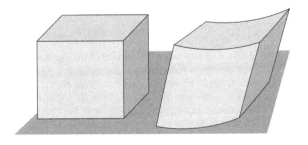

be done is to find the γ factor, that is at which speed test masses will vertically free-fall at a certain distance from the center. We will do so in Chap. 8 and get the **Schwarzschild exact solution**. The **Newton law of gravity** will follow from it for weak gravity. This case is most important because stars and planets are perfect balls to a considerable degree.

2. Observations show that over large enough distances, there is more or less the same amount of mass *everywhere* in the universe. Then again mass is balanced because the bending space-time must be the same everywhere in *space* and can only depend on *time*. We thus will see in Sect. 9.8 how the big-bang of the universe comes about.

And basically, that's that! There are some slightly more general solutions for a perfectly rotating ball, or a perfect ball with electric charges, or even a rotating perfect ball with electric charges, and the like. However, for an *arbitrary* collection of masses, we have to use tensor analysis and have to solve approximately the Einstein equation of gravity on a computer.

Make no mistake, we cannot *prove* that the law of gravity must be the Einstein equation. We only said that the Einstein gravity law is the *simplest possible* one. Physicists constructed other theories in which mass changes the space-time around it in a more complicated way. However, then it becomes more and more difficult to accommodate the law of gravity to the equivalence principle. In other words: it is hard to construct a theory which correctly describes how mass *creates* gravity *and* how mass *reacts* on gravity. What is more, experiments and observations showed again and again that only the Einstein gravity law plus the equivalence principle seem to lead to the correct law of gravity.

Now look how beautifully everything falls into place:

1. We fixed positions and speeds for test masses inside a small cloud such that they rest relative to each other and then let them loose. The simplest way mass can bend space-time is the *rate* at which the volume *begins* to shrink, while *free*-falling.
2. This shrinking rate does *only* depend on the mass *inside* the volume.
3. It does so in the simplest manner, that is in proportion to the mass inside.
4. The gravity law fits in with the theory of special relativity because it does not change even if we free-fall relative to the free-falling cloud.
5. *One physical* quantity, that is to say the mass *inside* the cloud, determines how another *geometrical* quantity, that is to say its *volume*, changes with time. In other words: the Einstein equation of gravity does *not* regulate directly how the *form* of the small volume changes with time!
6. This is done by masses *outside* the cloud. They *deform* the cloud but do not change the *volume* to fit the global picture of the bending space-time.

The theory of using the Einstein equivalence principle and the Einstein gravity law is the **theory of general relativity**.

Next we want to see the gravity law in action. How does the motion of the planets around the sun follow from it? How does this law fit in with the classic Newton way of looking at gravity, where the sun seems to "pull" Earth around it? And what new effects can we find?

Chapter 8
Solving the Einstein Equation of Gravity

Having stated and discussed the Einstein equation of gravity, let us solve them!

8.1 Gravity Causes the Law of Motion

The equivalence principle told us that test masses free-fall under gravity, that is how they *react* on bended space-time. This is the **law of motion**. The Einstein equation of gravity tells us how mass *acts* on space-time, how it bends space-time. Surprisingly the Einstein equation of gravity also tells us how test masses must *react*! A thought experiment will tell us the why. Suppose that we rest inside a small cloud of test masses in an inertial state. The test masses are also resting relative to us, as sketched in Fig. 8.1.

According to the Einstein equation of gravity, the cloud will begin to shrink at a rate which is in proportion to the mass inside the cloud. Now let us look at an even smaller volume inside the cloud, directly around us. The smaller this volume is, the less mass is inside. The less mass is inside, the less the volume will begin to shrink.

What is more, the cloud can begin to change its form as in Fig. 5.12. This is due to the gravity of outside masses. However, compared with the planet nearby, the volume we showed there is quite large. Again, the smaller volume we identify around us, the less it will begin to deform.

Therefore let us choose a tiny volume inside the cloud which contains only the white test mass beside us and us. This volume will neither begin to shrink nor begin to change its form. However, this is only possible if the white test mass beside us will *not* begin to move relative to us, that is it will not *accelerate* relative to us. That means that the white test mass is also in an inertial state, moving along a geodesic in space-time.

In other words: we have found the law of motion for this white test mass!

Compare this with the Newton theory of gravity: the gravity of Earth creates a force pulling at a test mass. This is the Newton law of gravity telling us how matter *creates* gravity. Then the test mass *reacts* by resisting the acceleration because of

© Springer International Publishing Switzerland 2015
K. Fischer, *Relativity for Everyone*, Undergraduate Lecture Notes
in Physics, DOI 10.1007/978-3-319-17891-2_8

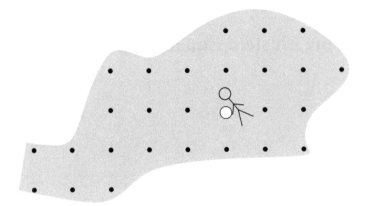

Fig. 8.1 The *white* test mass directly beside us will not begin to accelerate relative to us

its *inertia*. We cannot infer this reaction of the test mass from the Newton law of gravity. What is more, it is a riddle why weight and inertia should be equal at all. In fact, the Newton law of gravity is only an estimate of what really happens.

Compare gravity with electrodynamics: obeying the Maxwell equations, electric charges create an electromagnetic field around them. But this fact *alone* does not tell us how another charged mass will *react* to the field. We know that the **Lorentz force** of Chap. 3 describing this reaction fits in with both the theory of relativity and the Maxwell equations. However, the Lorentz force is not the only *possible* force fitting in with the Maxwell equations.

Incidentally, an electromagnetic field carries energy and hence bends space-time so that using the Einstein equation of gravity and a similar argument as above, we can derive the Lorentz force!

8.2 Gravity Inside a Perfect Ball of Mass

A star or a planet is nearly a perfect ball. In a thought experiment, we drill thin vertical shafts into the ball through its center, in many directions. For a better view, we sketched only four directions in Fig. 8.2. We place test masses in the shafts at the same fixed distance from the center, which are touching the small black ball in the center. At some time we release the test masses and let them loose so that they now fall freely.

If the black volume is small enough and if the test masses rest relative to each other when we release them, we can use the Einstein equation of gravity for it: the rate at which the black volume marked by the free-falling test masses *begins* to shrink is in proportion to the mass inside, that is the mass of the black ball. Because the mass spreads in all directions in the same way, the test masses all fall inside the vertical shafts at the same speed and mark a ball at all times.

Fig. 8.2 The *small black* disks at the surface of the *central black mass* are the test masses which start to free-fall into vertical shafts of the perfect ball. The *dashed arrow* marks us free-falling into one of the shafts and having started from a position where we were resting far away enough from the perfect ball

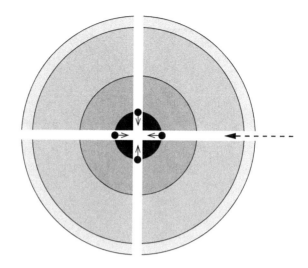

Make no mistake. The mass of the perfect ball *itself* does *not* move at all! Only the *test masses* begin to move because of the gravity of the *motionless* mass *inside the perfect ball*.

We now measure the space-time inside and around the perfect ball. However, we know from Sects. 6.1 and 6.4 that under gravity time itself runs different for different distances to the center and that vertical lengths also change under gravity. We also know from Sect. 6.6 that even the mass of a section of the perfect ball itself changes by the gravity of other parts of the perfect ball.

Using the equivalence principle, we account for it as follows. We start from a point at the right, nearly resting in an inertial state far away relative to the perfect ball, and free-fall from the right. In Fig. 8.2, we sketched this by the horizontal dashed arrow. We enter the shaft and at the time when we pass the center, all the test masses begin to free-fall. Because we are free-falling, our proper time is the same time as if resting far outside the perfect ball.

We know from Sect. 7.4 that the Einstein equation of gravity does not change if we measure mass and volume of the small ball marked by the test masses, as well as time, from our point of view.

Next, we repeat this thought experiment with a small black volume sandwiched between the ball at the center and some slightly larger ball with the same center, as in Fig. 8.3. The shell between the two balls should be very thin. We only drew it thicker for a better view.

Again we free-fall from our initial resting place far outside. Again the volume marked by the test masses is very small. Again the test masses do not move relative to each other. Again the test masses begin their free-fall when we pass through that volume. Therefore, we can use the Einstein equation of gravity. The volume marked by the test masses begins to shrink in proportion to the mass in the black volume. The same is true for any other small volume inside this shell in any other direction.

Fig. 8.3 A small, thin *black* volume sandwiched between two balls begins to shrink

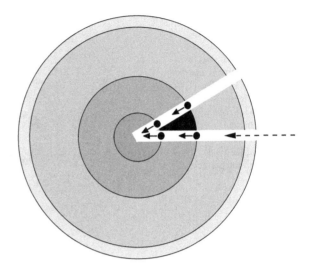

Hence it is true for the total thin shell sketched as black region in Fig. 8.4. Again we drew the thin shell not so thin for a better view.

In both Figs. 8.2 and 8.3 we free-fall. The clock with which we measure the beginning of the shrinking is proceeding at the same pace as for bodies resting far outside the perfect ball. Hence, if we let loose the test-masses in Fig. 8.5 at the same time, the *total* ball marked by the masses on the outside of the thin shell also begins to shrink in proportion to *its* mass. This total mass is the black volume in Fig. 8.5.

Fig. 8.4 A thin *black* shell inside the perfect ball. We drew the shell not so thin for a better view, and did not draw the test masses sitting at its *inner* and *outer* surface

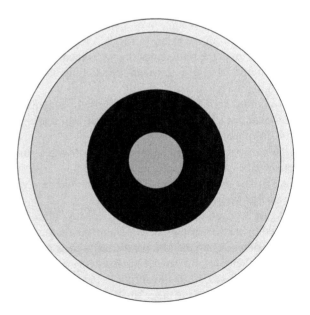

Fig. 8.5 A slightly larger
ball marked by test masses
begins to shrink

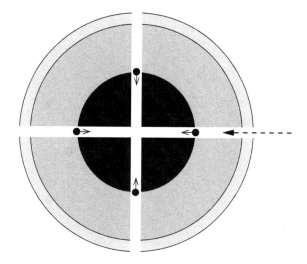

Thus we mark piece by piece towards the outside by adjoining thin shells like an onion skin. Repeating the above-mentioned argument, we conclude that the volume of the total perfect ball marked by test masses begins to shrink at a rate in proportion to the mass inside the ball.

8.3 Flat Space-Time Inside a Ball-Shaped Hollow

Now let us think of a weird planet that has a hollow space in its center, as sketched in Fig. 8.6. We can use the arguments of the last section for the ball-shaped hollow itself. Place test masses at the rim of the hollow. Then the ball marked by these test masses will begin to shrink in proportion to the mass it contains, that is zero. In other words: it will not begin to shrink at all. However, it also must remain ball-shaped because the mass on the outside spreads equally in all directions. Hence the test masses do not begin to move at all. Inside the hollow there is *no* gravity.

This is our first **exact solution** of the **Einstein equation of gravity**! It is called the **Birkhoff theorem**:

Birkhoff Theorem
If a **perfect ball** contains a ball-shaped **hollow** with the same center as the ball, then inside that hollow **space-time does not bend**.

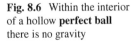

Fig. 8.6 Within the interior of a hollow **perfect ball** there is no gravity

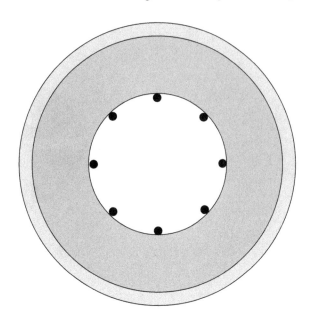

8.4 Gravity Outside a Perfect Ball of Mass

We continue with the argument of Sect. 8.2 as to the outside of the perfect ball and all that by adding thin shells piece by piece also on the outside of the perfect ball. **Outside the perfect ball** there is no more mass so that we conclude,

> The volume of any ball with the same center as the perfect ball—but larger than it and marked by test masses resting relative to each other—will begin to shrink at a rate which is 4π times the gravity constant times the mass of the perfect ball.
>
> We measure time and lengths free-falling from a resting place far away from the perfect ball.

Let us put this differently, namely in terms of the *radius* of such a ball, as sketched in Fig. 8.7. The test masses rest relative to each other and mark the volume of the larger of the dashed balls outside the perfect ball. At some time, they are released and begin to free-fall. We started free-falling from a resting place far away from the perfect ball and pass, say, the right test mass just as it begins to free-fall. During a short time the larger dashed ball shrinks a bit to become the smaller dashed ball. We saw in Sect. 6.5, point 6, that the distance of the two balls is then for us just the amount that the radius has shrunk during that short time.

Fig. 8.7 The marked ball around the perfect ball begins to shrink, as we pass it. In the figure, we sketched our passing by the *dashed arrow*

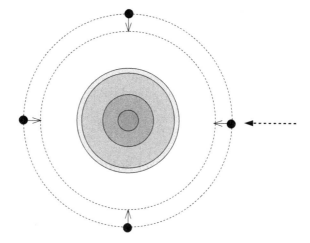

When the larger dashed ball begins to shrink to the slightly smaller dashed ball, it loses the volume which is between the two balls. As you see from Fig. 8.7 and as we said above, this is nearly the surface of the larger ball, times the distance between the two balls, or the difference of the radiuses of the two balls. The surface area of a ball with some radius is the same as in school geometry, that is 4π times the square of the radius, as we saw in Sect. 6.5, point 5. Hence the rate at which the volume of the dashed ball begins to *shrink* is nothing but the rate at which the radius begins to shrink, times 4π times the square of the radius.

In other words: according to the Einstein equation of gravity,

$$4\pi \times (\text{radius})^2 \times (\text{acceleration of the radius})$$
$$= -4\pi \times (\text{gravity constant}) \times (\text{mass of the perfect ball})$$

We need the "minus" because the radius is shrinking, so its acceleration is negative. The factor 4π drops out and we divide by the squared radius so that the change of the radius accelerates at a rate of

$$\begin{pmatrix} \text{acceleration of radius} \\ \text{in proper time of observer} \\ \text{free-falling from resting place} \\ \text{far away from perfect ball} \end{pmatrix} = -\frac{\begin{pmatrix} \text{gravity} \\ \text{constant} \end{pmatrix} \times \begin{pmatrix} \text{mass of} \\ \text{perfect ball} \end{pmatrix}}{(\text{radius})^2}$$

$$(8.1)$$

Even if our friend passes us at the given radius, free-falling with *less speed*, the equivalence principle tells us that we and our friend are both in an inertial state so that we do *not* accelerate relative to each other. Hence for a given radius,

Eq. (8.1) remains valid for a free-falling test mass of *any* possible speed relative to the perfect ball.

We see that although the change of the radius accelerates, the observer does *not* feel any acceleration. Again, this is because space-time *bends*. In the same way, an astronaut in the spaceship of Fig. 5.4 does not feel any acceleration, despite going along a bended path around Earth. This **acceleration** is only **relative** to the perfect ball.

8.5 Schwarzschild Exact Solution

We let a test mass free-fall from a place resting far away from the perfect ball. Let us calculate at which speed the test mass is approaching the perfect ball to complete the picture of Sect. 6.5. The test mass starts far away from the perfect ball at zero speed, relative to the perfect ball. The rate at which the speed is growing is the acceleration. For any radius we know this from Eq. (8.1). Hence we can calculate the γ factor for any radius, bit by bit. After what we learned in Sect. 6.4, this is enough to know how space-time bends. First we tell the result for the SPEED, at a given RADIUS,

$$(\text{SPEED})^2 = \frac{2 \times \left(\begin{array}{c}\text{gravity}\\\text{constant}\end{array}\right) \times \left(\begin{array}{c}\text{mass of}\\\text{perfect ball}\end{array}\right)}{(\text{RADIUS})} \qquad (8.2)$$

Let us now go backwards and confirm that this SPEED must have come from the **acceleration of the RADIUS** of Eq. (8.1). First we see from Eq. (8.2) that for a very large RADIUS, the SPEED becomes very small so that we really start to free-fall at nearly zero SPEED, if we rested only far enough away from the perfect ball. Next, we study the free-fall during a short time. We measure this time as our proper time. After the short time, the SPEED will grow by a small additional amount. During that short time, the RADIUS shrinks by a small amount. For us free-falling, this is just the small distance we traveled during that short time as we saw in Sect. 6.5, point 6. Hence, at this later time we have the equation

$$\left(\text{SPEED} + \begin{array}{c}\text{small additional}\\\text{speed}\end{array}\right)^2 = \frac{2 \times \left(\begin{array}{c}\text{gravity}\\\text{constant}\end{array}\right) \times \left(\begin{array}{c}\text{mass of}\\\text{perfect ball}\end{array}\right)}{\left(\text{RADIUS} - \begin{array}{c}\text{small}\\\text{distance}\end{array}\right)} \qquad (8.3)$$

The small additional speed per short time is the acceleration we search for. We expand the square on the left-hand side,

$$(\text{SPEED})^2 + 2 \times (\text{SPEED}) \times \left(\begin{array}{c}\text{small additional}\\\text{speed}\end{array}\right) + \left(\begin{array}{c}\text{small additional}\\\text{speed}\end{array}\right)^2$$

$$= \frac{2 \times \left(\begin{array}{c}\text{gravity}\\\text{constant}\end{array}\right) \times \left(\begin{array}{c}\text{mass of}\\\text{perfect ball}\end{array}\right)}{\left(\text{RADIUS} - \begin{array}{c}\text{small}\\\text{distance}\end{array}\right)}$$

The term in the middle on the left-hand side being the product of the original and the small additional speed is much larger than the product of the small additional speed with itself on the right. Hence we can neglect this term if the short time is short enough and we have

$$(\text{SPEED})^2 + 2 \times (\text{SPEED}) \times \left(\begin{array}{c}\text{small additional}\\\text{speed}\end{array}\right)$$

$$= \frac{2 \times \left(\begin{array}{c}\text{gravity}\\\text{constant}\end{array}\right) \times \left(\begin{array}{c}\text{mass of}\\\text{perfect ball}\end{array}\right)}{\left(\text{RADIUS} - \begin{array}{c}\text{small}\\\text{distance}\end{array}\right)}$$

For the squared SPEED on the left-hand side we replace the right-hand side of Eq. (8.2).

$$\frac{2\!\!\!\!/ \times \left(\begin{array}{c}\text{gravity}\\\text{constant}\end{array}\right) \times \left(\begin{array}{c}\text{mass of}\\\text{perfect ball}\end{array}\right)}{(\text{RADIUS})} + 2\!\!\!\!/ \times (\text{SPEED}) \times \left(\begin{array}{c}\text{small}\\\text{additional}\\\text{speed}\end{array}\right)$$

$$= \frac{2\!\!\!\!/ \times \left(\begin{array}{c}\text{gravity}\\\text{constant}\end{array}\right) \times \left(\begin{array}{c}\text{mass of}\\\text{perfect ball}\end{array}\right)}{\left(\text{RADIUS} - \begin{array}{c}\text{small}\\\text{distance}\end{array}\right)}$$

We cancel the factor 2 on both sides. In order to get the small additional speed, we subtract now the first term on both sides,

$$(\text{SPEED}) \times \left(\begin{array}{c}\text{small additional}\\\text{speed}\end{array}\right)$$

$$= \frac{\left(\begin{array}{c}\text{gravity}\\\text{constant}\end{array}\right) \times \left(\begin{array}{c}\text{mass of}\\\text{perfect ball}\end{array}\right)}{\left(\text{RADIUS} - \begin{array}{c}\text{small}\\\text{distance}\end{array}\right)} - \frac{\left(\begin{array}{c}\text{gravity}\\\text{constant}\end{array}\right) \times \left(\begin{array}{c}\text{mass of}\\\text{perfect ball}\end{array}\right)}{(\text{RADIUS})} \qquad (8.4)$$

If for example the original RADIUS was 100,000 m and we approached the perfect ball during the short time by the small distance of 3 meters, then the difference of the *inverse* radiuses is

$$\frac{1}{99,997} - \frac{1}{100,000} = \frac{100,000}{99,997 \times 100,000} - \frac{99,997}{99,997 \times 100,000}$$

$$= \frac{3}{99,997 \times 100,000} \approx \frac{3}{100,000^2}$$

Hence this is roughly the small distance divided by the *square* of the RADIUS. Multiplying this with $\left(\begin{array}{c}\text{gravity}\\\text{constant}\end{array}\right) \times \left(\begin{array}{c}\text{mass of}\\\text{perfect ball}\end{array}\right)$, we can rewrite the right-hand side of Eq. (8.4) as,

$$(\text{SPEED}) \times \left(\begin{array}{c}\text{small additional}\\\text{speed}\end{array}\right) = \left(\begin{array}{c}\text{gravity}\\\text{constant}\end{array}\right) \times \left(\begin{array}{c}\text{mass of}\\\text{perfect ball}\end{array}\right) \times \frac{\left(\begin{array}{c}\text{small}\\\text{distance}\end{array}\right)}{(\text{RADIUS})^2}$$

Now, the small additional speed per short proper time is the *acceleration* and the small distance per short time is the *negative* SPEED because the radius shrinks by that amount. Hence per short time we have

$$(\text{SPEED}) \times (\text{acceleration}) = -\left(\begin{array}{c}\text{gravity}\\\text{constant}\end{array}\right) \times \left(\begin{array}{c}\text{mass of}\\\text{perfect ball}\end{array}\right) \frac{(\text{SPEED})}{(\text{RADIUS})^2}$$

We divide both sides of the equation by the SPEED and end up with the Eq. (8.1). This shows us that Eq. (8.2) really gives the correct SPEED of a test mass free-falling from a place resting far away from the perfect ball. From it we get the γ factor,

$$\gamma = \sqrt{1 - \frac{\text{speed}^2}{c^2}} = \sqrt{1 - \frac{2 \times \left(\begin{array}{c}\text{gravity}\\\text{constant}\end{array}\right) \times \left(\begin{array}{c}\text{mass of}\\\text{perfect ball}\end{array}\right)}{(\text{radius}) \times c^2}} \qquad (8.5)$$

Hence we have **exactly solved** the **Einstein equation of gravity**, that is we derived the **Schwarzschild exact solution**:

Schwarzschild Exact Solution

1. A **perfect ball** of mass bends space-time such that at a certain radius time is running more slowly by the factor γ of Eq. (8.5), which is relative to a clock resting far enough away from the perfect ball.
2. The vertical distance between two nearby points will be *larger* than the difference of their radiuses, by the factor $1/\gamma$. The radius is therefore *smaller* than the distance to the center of the perfect ball.
3. Horizontal lengths do not change.
4. The ratio of the perimeter and radius of any circle with the same center as the perfect ball is 2π, and the surface area of a ball with the same center as the perfect ball is 4π times the *squared* radius.

8.6 Newton Law of Gravity

The Newton law of gravity follows from the Einstein gravity law, if the gravitating mass is small. Hence test masses resting far away from the gravitating mass will not reach large speeds while free-falling towards it. For example, according to Table A.1, the radius of the sun is 7×10^8 and its mass is 2×10^{30}. The gravity constant is about 7×10^{-11}. According to Eq. (8.2) we find that such a test mass free-falling from a resting place far away into the sun reaches at most 0.2 percent of the speed of light,

$$\frac{\text{speed}}{c} = \sqrt{\frac{2 \times (\text{gravity constant}) \times \text{mass}}{(\text{radius}) \times c^2}}$$

Let us insert the above numbers:

$$\frac{\text{speed}}{c} \approx \sqrt{\frac{2 \times \left(7 \times 10^{-11}\right) \times \left(2 \times 10^{30}\right)}{\left(7 \cdot 10^8\right) \times \left(3 \cdot 10^8\right)^2}} \approx \sqrt{\frac{2^2}{10^6} \times \frac{10}{9}} \approx \frac{2}{1000} \quad (8.6)$$

Then the γ factor is nearly one,

$$\gamma = \sqrt{1 - \frac{\text{speed}^2}{c^2}} \approx 0.999998$$

Therefore time runs everywhere nearly at the same pace and the length of the same rod is nearly everywhere the same. The radius is now also for an observer who is resting near the sun nearly the distance from the center. School geometry is nearly valid. The radius entering the Eq. (8.1) is now practically the distance of the test mass to the center. Hence the Einstein equation of gravity tells us that a test mass falling vertically towards a perfect ball will **accelerate relative** to the perfect ball at a rate

$$
\begin{pmatrix} \text{acceleration of} \\ \text{distance to center} \end{pmatrix} = - \frac{\begin{pmatrix} \text{gravity} \\ \text{constant} \end{pmatrix} \times \begin{pmatrix} \text{mass of} \\ \text{perfect ball} \end{pmatrix}}{(\text{distance to center})^2} \tag{8.7}
$$

This already looks very much like the Newton law of gravity. In fact, this law merges *two* laws of the original Newtonian mechanics. The first law states that there is a force coming somehow from the center of the perfect ball, acting infinitely fast at any distance, pulling a test mass of some heaviness towards the center of the perfect ball in proportion to its heaviness,

$$
\begin{pmatrix} \text{force on} \\ \text{test mass} \end{pmatrix} = - \frac{\begin{pmatrix} \text{gravity} \\ \text{constant} \end{pmatrix} \times \begin{pmatrix} \text{mass of} \\ \text{perfect ball} \end{pmatrix}}{(\text{distance to center})^2} \times \begin{pmatrix} \text{heaviness of} \\ \text{test mass} \end{pmatrix} \tag{8.8}
$$

This is the **Newton law of gravity**. At the same time, the inertia of the test mass resists the force such that the test mass accelerates in inverse proportion to its inertia

$$
\text{acceleration} = \frac{\begin{pmatrix} \text{force on} \\ \text{test mass} \end{pmatrix}}{\begin{pmatrix} \text{inertia of} \\ \text{test mass} \end{pmatrix}} \tag{8.9}
$$

This is the **Newton law of motion**. Dividing both sides of the Newton law of gravity (8.8) by the "inertia of test mass", we see how the acceleration depends on the distance to the center of the perfect ball:

$$
\text{acceleration} = \frac{\begin{pmatrix} \text{force on} \\ \text{test mass} \end{pmatrix}}{\begin{pmatrix} \text{inertia of} \\ \text{test mass} \end{pmatrix}} = - \frac{\begin{pmatrix} \text{gravity} \\ \text{constant} \end{pmatrix} \times \begin{pmatrix} \text{mass of} \\ \text{perfect ball} \end{pmatrix}}{(\text{distance to center})^2} \times \frac{\begin{pmatrix} \text{heaviness of} \\ \text{test mass} \end{pmatrix}}{\begin{pmatrix} \text{inertia of} \\ \text{test mass} \end{pmatrix}}
$$

In the Newton theory, the heaviness and the inertia of a test mass are *accidentally* equal so that they cancel each other on the right-hand side of the equation, and we have again the law (8.7).

However, there is no *reason* why inertia and heaviness should be the same within the Newtonian mechanics. We see that in general relativity, inertia and heaviness are *equal because* gravity is *no* force. All kinds of test masses move *in the same way* of free-falling through space-time bended by masses. We also see how the classic Newton law of gravity *and* the Newton law of motion emerge as a good estimate of the Einstein equation of gravity, if the gravitating masses are not too large! The Newton law of motion emerges because of what we said in Sect. 8.1 that the Einstein equation of gravity also determines the law of motion of a test mass.

Chapter 9
General Relativity in Action

9.1 Black Holes

A nearly perfect ball such as the sun has a mass of 2×10^{30} and a radius of 7×10^8. These two numbers determine how *strongly* this star bends space-time via the following ratio

$$S = \frac{2 \times \left(\begin{array}{c} \text{gravity} \\ \text{constant} \end{array} \right) \times \left(\begin{array}{c} \text{mass of} \\ \text{perfect ball} \end{array} \right)}{c^2} \qquad (9.1)$$

The letter **S** stands for the **Schwarzschild radius** of the perfect ball. For the sun, we have in meters

$$S = \frac{2 \times \left(6.67 \times 10^{-11}\right) \times \left(2 \times 10^{30}\right)}{\left(3 \times 10^8\right)^2}$$
$$\approx 2.96 \times 10^3 \approx 3000 \, \text{m} \qquad (9.2)$$

which is much smaller than its radius of about $7 \times 10^8 \, \text{m}$. The vertical speed of Eq. (8.2) at which a test mass resting far away from a perfect ball free-falls towards the perfect ball becomes in terms of the Schwarzschild radius

$$\frac{\text{speed}^2}{c^2} = \frac{\dfrac{2 \times \left(\begin{array}{c} \text{gravity} \\ \text{constant} \end{array} \right) \times \left(\begin{array}{c} \text{mass of} \\ \text{perfect ball} \end{array} \right)}{c^2}}{\text{radius}} = \frac{S}{\text{radius}} \qquad (9.3)$$

© Springer International Publishing Switzerland 2015
K. Fischer, *Relativity for Everyone*, Undergraduate Lecture Notes
in Physics, DOI 10.1007/978-3-319-17891-2_9

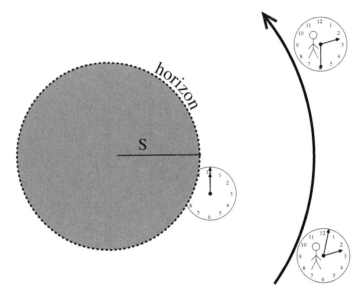

Fig. 9.1 Clocks run more slowly near large masses and freeze at the horizon of a black hole relative to outside observers

The Schwarzschild radius enters the γ factor of the Schwarzschild exact solution (8.5) as

$$\gamma = \sqrt{1 - \frac{2\left(\begin{array}{c}\text{gravity}\\\text{constant}\end{array}\right) \times \left(\begin{array}{c}\text{mass of}\\\text{perfect ball}\end{array}\right)}{(\text{radius}) \times c^2}} = \sqrt{1 - \frac{S}{\text{radius}}} \qquad (9.4)$$

which is for the sun about

$$\gamma \approx 0.999998$$

Please have a look at Fig. 9.1. We know from the Schwarzschild solution that compared to a distant resting observer, a clock of an observer on the surface of the ball runs more slowly by the factor γ. For the same radius, the larger the mass of the ball is, the larger the ratio $\dfrac{S}{\text{radius}}$ is, and the stronger gravity acts on the surface of the perfect ball.

When the mass in the perfect ball clumps so much that its radius is *smaller* than its Schwarzschild radius, then a strange thing happens. A clock at the Schwarzschild radius is still outside the perfect ball with a γ factor of

$$\gamma = \sqrt{1 - \frac{S}{S}} = 0$$

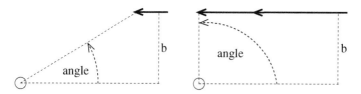

Fig. 9.2 Light passes near a light perfect ball nearly along a straight line

That is, an outside observer sees that the time of the clock is *frozen*! Such a star is a **black hole**. We already discussed this in Sect. 6.8, but here we see how the Schwarzschild exact solution predicts that black holes are possible. The **horizon** which we discussed in Sect. 6.8 is therefore the surface of the ball which has its center at the black hole, and its radius is the Schwarzschild radius.

Outside the black hole, its gravity acts like any other perfect ball. Equation (9.2) tells us that if the sun was able to shrink to less than about 3000 m it would become a black hole. However, the planets would move exactly on the same paths around the sun as they do now!

In principle the theory allows *any* body to become a black hole, if only it shrinks enough. Of course, this depends on the inner structure of the body. Indeed, you cannot squeeze a stone without much effort. For collapsing stars this depends on how rigid its elementary particles are. With the help of **quantum theory**, we can predict that a star must have at least roughly one and a half times the mass of the sun to be able to collapse eventually into a **black hole**.[1] Up to now (2015), the smallest *observed* black holes have a mass of about three times the mass of the sun.[2] Large black holes should sit in the center of a galaxy, "feeding" on stars which came to close. It seems that in the center of our galaxy there is a monster black hole which has a mass of over four million suns.[3]

9.2 Light Bending: Weak Gravity 1

In Fig. 6.10 we already saw that a light beam should *bend* near a large mass. How much does a light beam bend?

A light beam passes a perfect ball from the right to the left, as sketched in Fig. 9.2. The light beam is the solid black arrow and all other dashed lines are only a guide for the eye. The gray circle at the bottom of the figure shows a very light perfect ball. We do not aim at the center of the ball but at some distance *b* from it. Because the ball

---

[1] See for example http://en.wikipedia.org/wiki/Tolman-Oppenheimer-Volkoff_limit and http://en.wikipedia.org/wiki/Chandrasekhar_limit.

[2] See for example http://www.nasa.gov/topics/universe/features/black-hole-heartbeat.html or http://en.wikipedia.org/wiki/IGR_J17091-3624.

[3] See for example http://en.wikipedia.org/wiki/Black_hole.

Fig. 9.3 The total angle
between the incoming and
the outgoing direction will
grow eventually to 180
degrees

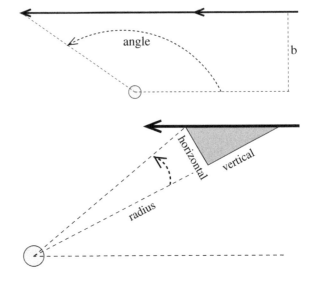

Fig. 9.4 The *dashed arrow*
shows the growth of the
angle during some short time
like, say, one millisecond.
We draw the *gray triangle*
much larger than in reality
for a better view

is very light, it nearly has no gravity. Hence the light beam will pass along a straight
line. We measure how far the light has gone by the angle it has with the dashed line
that runs from the left to the right. At the far right, the angle is practically zero. It
grows as the light beam approaches the perfect ball. When it is nearest, the angle is
90 degrees, as in the right-hand picture of Fig. 9.2. When the light has passed far to
the left, the angle grows eventually to 180 degrees, as you can see in Fig. 9.3.

First we need to know at which rate the angle is growing when a very light perfect
ball is present. In Fig. 9.4 we see the light passing during some short time, say one
millisecond. During this time the angle grows as much as the dashed arrow shows.

How much does the angle grow? The ratio of the growth and the full angle of 360
degrees is the ratio of the horizontal side of the gray triangle and the perimeter of
the circle with the radius,

$$\frac{\text{angle growth}}{360 \text{ degrees}} = \frac{\text{horizontal}}{2\pi \times \text{radius}} \tag{9.5}$$

Beware: "horizontal" does *not* mean "along the lines of the text in the book" but "at
right angles to the respective radius"! Next, we replace the light by a heavy perfect
ball. Then as the light passes it to the left, the angle will grow *beyond* 180 degrees, as
you see in Fig. 9.5: the light beam *bends*. We estimate now this additional **bending
angle of the light beam** by using the Schwarzschild exact solution.

The Schwarzschild solution Sect. 8.5, point 2, tells us that the small gray triangle
in Fig. 9.4 will distort because space is bending: the horizontal side will not change
but the vertical side will grow by the $1/\gamma$ factor of the radius at which the triangle

Fig. 9.5 A light beam bends near a heavy perfect ball

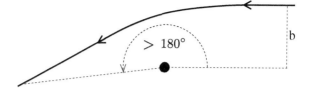

exists. We flatten out space by shrinking the vertical side and the radius by the same amount,

$$\text{radius} \longrightarrow \gamma \times \text{radius} \tag{9.6}$$

Time also "bends". The light beam passes along the longest side of the small triangle at the speed c. However, we observe the light from a place far away "above" the perfect ball, resting relative to the perfect ball in an inertial state. Relative to us, time slows down by the factor γ.

$$\text{short time} \longrightarrow \gamma \times (\text{short time}) \tag{9.7}$$

The ratio of the horizontal side per time, per radius, is in proportion to how much the angle will grow per short time,

$$\frac{\text{horizontal}}{(\text{short time}) \times \text{radius}} \longrightarrow \frac{\text{horizontal}}{(\gamma \times \text{short time}) \times (\gamma \times \text{radius})}$$
$$= \frac{\text{horizontal}}{(\text{short time}) \times (\gamma^2 \times \text{radius})} \tag{9.8}$$

We see: if the perfect ball is heavy, the angle grows per short time interval by an additional factor γ^{-2}. In other words: the light beam bends. However, then Fig. 9.4 cannot be correct because the light will not follow a straight line! Nonetheless we can use this figure if the perfect ball is not too heavy, as for example the sun. Then the light beam bends only a little. Hence it does not make much difference if we assume that the light beam came in along a straight line, from the right. We make only a small mistake by adding up these *bending angles* for each short time. In other words: once we have flattened out space-time near the place where the light beam passed, we can use school geometry to calculate the *bending* angle. Physicists call it a **perturbative calculation**.

The second line of Eq. (9.8) tells us that, instead of changing the horizontal side of the triangle and time, we also can *shrink* the radius by the amount γ^2 and leave all other distances and times as they are. In this way, we can flatten space-time around a not too heavy perfect ball with sufficient accuracy.

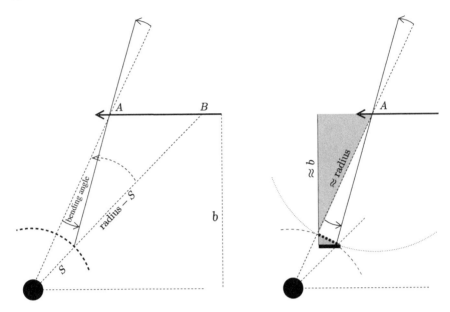

Fig. 9.6 Light bends because it feels only the radius minus the Schwarzschild radius. For a better view, we drew the Schwarzschild radius S larger than the radius of the black perfect ball. For a not too heavy perfect ball, the Schwarzschild radius is much smaller than either b or the radius. Hence to a very good approximation, the left side of the *larger gray triangle* has length b, and its longest side is nearly the radius

We know from the Schwarzschild exact solution (9.4) that the square of the γ factor is $\left(1 - \dfrac{S}{\text{radius}}\right)$. Hence we shrink the radius to

$$\text{radius} \longrightarrow \gamma^2 \times (\text{radius}) = \left(1 - \frac{S}{\text{radius}}\right) \times (\text{radius}) = \text{radius} - S \qquad (9.9)$$

In words: in order to calculate the *bending* angle, we have to shrink the radius by the Schwarzschild radius.

Next, please have a look at the left-hand picture of Fig. 9.6. For a very light perfect ball, the light beam travels during a short time along the arrow from point B to point A. The additional angle during that short time is the dashed angle measured from the center of the perfect ball, that is in the same way as in Fig. 9.4.

For a heavier perfect ball we have to use the shrunken radius, "radius $- S$" of Eq. (9.9). That is, we measure the angle not from the center of the perfect ball but from the Schwarzschild radius. We get the thin solid line through A. This line turns against the left dashed line by the solid *bending* angle.

In the right-hand picture we show now that the *bending* angle grows in proportion to the **solid fat** side of the small gray triangle.

The **solid fat** side of the small gray triangle stands at right angles to the solid side b of the larger gray triangle. Its largest side is the **dashed fat** line which is nearly part of the dotted circle around A and is therefore standing at right angles to the radius which is the largest side of the larger gray triangle. Its smallest side stands at right angles to the smallest side of the larger gray triangle. Hence if we turn the small gray triangle clockwise by 90 degrees and enlarge it, we could fit the larger triangle. Therefore the sides of the small and the large gray triangle are in proportion. In particular,

$$\frac{\textbf{solid fat } \text{line}}{b} = \frac{\textbf{dashed fat } \text{line}}{\text{radius}} \tag{9.10}$$

The **dashed fat** line is also nearly that part of the dotted circle with center in A, belonging to the *bending* angle in the left-hand picture. Hence the ratio of the **dashed fat** line and the perimeter of $2\pi \times$ radius of the dotted circle equals the ratio of the *bending* angle and the full angle,

$$\frac{bending \text{ angle}}{360 \text{ degrees}} = \frac{\textbf{dashed fat } \text{line}}{2\pi \times \text{radius}}$$

We multiply both sides with 2π,

$$2\pi \times \frac{bending \text{ angle}}{360 \text{ degrees}} = \frac{\textbf{dashed fat } \text{ line}}{\text{radius}} \tag{9.11}$$

Because the right-hand sides of the two Eqs. (9.10) and (9.11) are equal, the left-hand sides are also equal. Indeed the *bending* angle is growing in proportion to the **solid fat** line,

$$\frac{\textbf{solid fat } \text{line}}{b} = 2\pi \times \frac{bending \text{ angle}}{360 \text{ degrees}}$$

Next have a look at Fig. 9.7. As the light beam travels from the right to the left, the **solid fat** line is growing all the way from the right to the left of the dashed circle, becoming eventually its *diameter* $2S$. Hence the total *bending* angle becomes

$$\frac{2S}{b} = 2\pi \frac{\text{total } bending \text{ angle}}{360 \text{ degrees}}$$

We cancel the factors 2. After multiplying both sides of the equation with $\dfrac{360}{\pi}$ degrees, we finally get how much a light beam is bending around a not too heavy perfect ball,

Fig. 9.7 The *bending* angle
grows in proportion to the
solid fat line

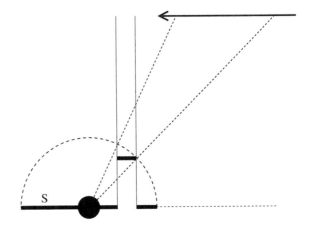

$$\frac{360}{\pi} \times \frac{S}{b} \text{ degrees} = \text{total } \textit{bending} \text{ angle} \qquad (9.12)$$

To get the largest effect for a given star as the sun, we should make b as small as possible. We should use light beams just touching the sun so that b is then the radius of the sun. Multiplying with 60 gives it in **arc minutes**, and again with 60 gives the bending angle in **arc seconds**, and filling in the numbers from Table A.1, gives

$$\text{bending angle} \approx \frac{360}{\pi} \times \frac{2.96 \times 10^3}{6.96 \times 10^8} \times 60 \times 60 \approx 1.75 \text{ arc seconds} \qquad (9.13)$$

This is what many experiments have shown over the years![4]

Observe that to get the correct bending angle, we needed three effects:

1. Space bends around the perfect ball.
2. Time slows down near the perfect ball.
3. A light beam passes near the observer at speed c.

If only space bended, as popular texts often depict, the above argument would then show that we would have to ignore Eq. (9.7), and have to shrink the radius in Eq. (9.8) not by the factor γ^2 but only by the factor γ. Hence our calculation would predict *a lesser bending* (in fact, about one-half) than the one observed. So it is really again space-*time* which is bending.

---

[4]See for example: http://en.wikipedia.org/wiki/Gravitational_lens or for the first observation: http://rsta.royalsocietypublishing.org/content/220/571-581/291.full.pdf.

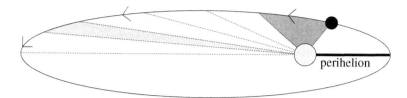

Fig. 9.8 The sun is the *light-gray ball* being in one focus point of the ellipse. The planet is the *black ball* on the ellipse. The **perihelion** is the line at which the planet comes nearest to the sun. The *dotted straight lines* show where the planet is after a fixed time span has passed. The *light gray* and the *darker gray triangles* have the same area so that the planet is faster near the sun than far away from it

9.3 Kepler Laws

We saw in Sect. 8.6 that for weak gravity the Newton law of gravity is nearly correct. The Newton law of gravity can explain the famous three **Kepler laws**. Planets going around the sun move approximately according to these three laws:

1. A planet moves around the sun on an **ellipse** with the sun sitting in one focus of the ellipse.
2. A straight line extending from the center of the sun to the center of the planet moves over equal areas during equal intervals of time. See Fig. 9.8.
3. The square of the annual period of the planet is in proportion to the third power of the *largest* radius of the ellipse. This largest radius is half the length of the horizontal diameter line in Fig. 9.8. The constant of proportion is $4\pi^2$ divided by the gravity constant and the mass of the sun.

Here we explain only the simplest case in which the planets move in ellipses that are nearly circles. This case still shows all the interesting physical effects. Then the "largest radius" is just "the" radius to the sun. The **third Kepler law** says in simplified terms:

> The square of the annual period of the planet is in proportion to the third power of its radius to the sun. The constant of proportion is $4\pi^2$ divided by the gravity constant and the mass of the sun.

Example: Earth is at distance of 1.5×10^{11} to the sun that has mass of 2×10^{30}, and the gravity constant is 6.67×10^{-11}. The gravity of the sun is not too strong so that we can use the distance of Earth to the sun as the radius. Hence one year should be about

$$\text{year} = \sqrt{4\pi^2 \frac{\left(1.5 \times 10^{11}\right)^3}{\left(6.67 \times 10^{-11}\right) \times \left(2 \times 10^{30}\right)}} \approx 3.16 \times 10^7 \text{ seconds}$$

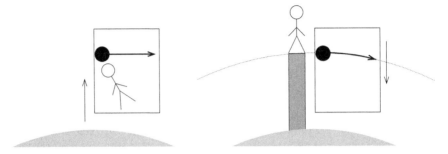

Fig. 9.9 A test mass moves free-falling in a circle around a perfect ball, if we tune its horizontal speed. We need no force for that, just the bended space-time

Check: including the leap years, one year is about 365.25 days, that are 365.25 × 24 hours or 365.25 × 24 × 60 × 60 ≈ 3.16 × 10⁷ seconds, as it should be.

How do the Kepler laws follow from the Einstein equation of gravity? First of all, the planets are much smaller than the sun, at least the rocky ones nearer to the sun. Therefore we assume that the planets themselves are just test masses. Then we know that the sun is nearly a perfect ball. How can such a test mass move around the perfect ball in a circle and at the same time free-fall? Please have a look at the left-hand picture of Fig. 9.9. The box free-falls vertically. Inside the box a test mass moves horizontally at a steady speed, as sketched by the horizontal thick arrow. The vertical arrow shows the perfect ball **accelerating relative** towards the box.

In the picture on the right-hand side, we stand on a platform that reaches up to the height where the test mass is. We see the box falling down faster and faster. Hence the test mass moves more and more downwards while moving to the right. It moves along a *bended* path, much in the same way as the light beam does in Fig. 6.10. If we adjust the horizontal speed, we can arrange that the bended path continues to be the dotted circular path around the perfect ball. Hence we see here how just by bending space-time, the sun makes a planet moving around it without any force pulling the planet. We show now how large this speed is by using the Einstein equation of gravity *exactly* so that we do *not* need the Newton law of gravity at all!

We look from a place far enough above the plane in which the planet moves, nearly resting relative to the perfect ball in an inertial state. We observe the planet during a short time interval, as in Fig. 9.10. During this short time it moves a certain distance along the circle to the right. This is nearly the horizontal distance in the left-hand picture of Fig. 9.10. Distance per time is speed, so it moves at a distance that is speed times the short time.

During this short time the place of the planet turns by a small angle. We learned in Sect. 6.5 that the perimeter of the circle with the same center as the perfect ball is 2π times the radius, even within bended space-time. Hence for a small angle, the ratio of horizontal distance and radius is in proportion to the angle.

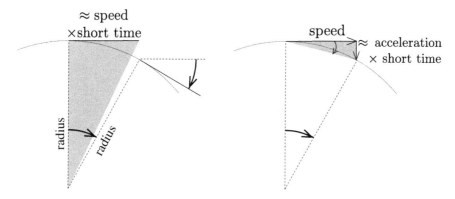

Fig. 9.10 During a short time the place and the speed of the planet turn at nearly the same angles so that the planet remains on the circle around the sun. We depicted the angles as *curved solid arrows*

Next let us have a look at how the speed is changing its direction during the short time, as in the right-hand picture. The additional speed is nearly vertical during the short time. Acceleration is additional speed per short time. So the additional speed is just the acceleration times the short time, as we see in the right-hand picture.

Now the decisive point is: to remain on the circle also in the future, the direction of the speed must turn by the *same* small angle during the short time. Because both the left and the right gray triangles have the same small angle and are nearly right triangles, the ratio of their smallest side and the side at right angles to it must be nearly the same. The smaller the time interval is, the more accurate it gets:

$$\frac{\text{speed} \times (\text{short time})}{\text{radius}} = \frac{\text{acceleration} \times (\text{short time})}{\text{speed}}$$

We can cancel the short time and we see how the speed depends on the acceleration and the radius:

$$\text{speed}^2 = (\text{acceleration}) \times (\text{radius})$$

The acceleration results from the Einstein equation of gravity (8.1). That means we are looking from a place far above the plane in which the planet moves and are resting relative to the perfect ball in an inertial state. We can ignore the minus sign in that equation because here we viewed the vertical speed in Fig. 9.10 as a positive number. Hence a planet moving along a circle with a certain radius moves at the speed

$$\text{speed}^2 = \frac{\left(\begin{array}{c}\text{gravity} \\ \text{constant}\end{array}\right) \times \left(\begin{array}{c}\text{mass of} \\ \text{perfect ball}\end{array}\right)}{(\text{radius})^2} \times (\text{radius})$$

We cancel the radius on the right against one radius in the denominator,

$$\text{speed}^2 = \frac{\left(\begin{array}{c}\text{gravity}\\\text{constant}\end{array}\right) \times \left(\begin{array}{c}\text{mass of}\\\text{perfect ball}\end{array}\right)}{(\text{radius})} \qquad (9.14)$$

In terms of the Schwarzschild radius (9.1) this is

$$\frac{\text{speed}^2}{c^2} = \frac{1}{2} \times \frac{2 \times \left(\begin{array}{c}\text{gravity}\\\text{constant}\end{array}\right) \times \left(\begin{array}{c}\text{mass of}\\\text{perfect ball}\end{array}\right)}{(\text{radius}) \times c^2}$$

$$\frac{\text{speed}^2}{c^2} = \frac{1}{2} \times \frac{S}{\text{radius}} \qquad (9.15)$$

Again this is an exact result of the theory of general relativity!

Now, speed is distance per time. During one year the test mass moves once around the circle. This distance is 2π times the radius, as we know from Sect. 6.5, point 5. Hence we can express the speed on the left-hand side of Eq. (9.14) in terms of the perimeter of the circle and duration of a year,

$$\frac{4\pi^2 \times (\text{radius})^2}{(\text{year})^2} = \frac{\left(\begin{array}{c}\text{gravity}\\\text{constant}\end{array}\right) \times \left(\begin{array}{c}\text{mass of}\\\text{perfect ball}\end{array}\right)}{\text{radius}}$$

Solving the length of the year, we get

$$\frac{4\pi^2 \times (\text{radius})^3}{\left(\begin{array}{c}\text{gravity}\\\text{constant}\end{array}\right) \times \left(\begin{array}{c}\text{mass of}\\\text{perfect ball}\end{array}\right)} = (\text{year})^2 \qquad (9.16)$$

This is just the **third Kepler law**. We see that if we use the radius as in Sect. 6.5 and measure the time from a point resting far away and relative to the perfect ball, then for a circular orbit, the third Kepler law remains *exactly* correct in the theory of general relativity.

9.4 Planet Orbits Rotate: Weak Gravity 2

In the previous section we stated the Kepler laws, in particular that the planets move around the sun on an ellipse. As a matter of fact, the planets do *not* follow an ellipse to perfection. The reason is mainly that the planets act on each other with their *own* gravity. As one result, the ellipse itself slightly deforms. This is hard to observe.

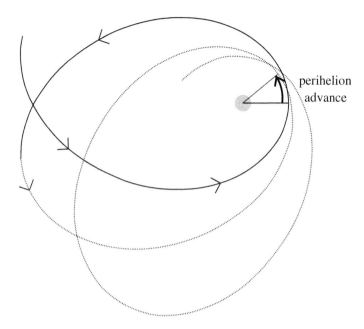

Fig. 9.11 A planet moves along a slowly rotating ellipse so that the orbit does not close properly. For a better view, we calculated the orbit of a planet which has roughly the same distance to the star as Earth has to the sun, but here the star is one-million times heavier than the sun

However, one effect *piles up* as the years go by so that astronomers can easily measure it, just by waiting long enough. The ellipses themselves *rotate* along the sun very slowly, as sketched in Fig. 9.11.

Astronomers use the line where the planet comes closest to the sun, as a reference line. It is called the **perihelion** and you see it in Fig. 9.8. Every time the planet comes nearest to the sun, this perihelion has turned at some angle. Astronomers call this angle per one turn the **perihelion advance**.

Mathematicians and physicists developed *calculus* among other things to *calculate* such tiny effects by using the Newton law of gravity. One way is to *deduct* the influence of the other planets. Then a lonely planet should move along a perfect ellipse because that is what the Newton law of gravity predicts for a single planet going around the nearly perfect ball of the sun.

However, what we really see is that the ellipse of such an imagined lonely planet *still* rotates by some angle per one turn around the sun! Now we know that the Newton law is out of line with the correct Einstein equation of gravity. So Einstein asked himself whether this left-over *perihelion advance* is an effect which can be explained by the general relativity.

First let us use the Einstein equation of gravity only in the approximation of the Newton law of gravity which leads to the Kepler laws. Then a test mass moves around a perfect ball on a fixed ellipse. So to make things simpler, assume that the ellipse differs from a circle only by a *tiny bit*.

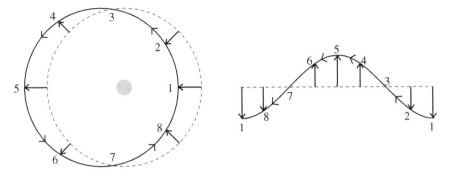

Fig. 9.12 *Left-hand picture* the perfect ball is at the center of the *dashed circle*. The *solid ellipse* is a circle to a very good approximation. The *arrows* show how much the test mass is departing from the circular orbit. Point 1 is the perihelion. *Right-hand picture* the *dashed circle* expanded to a *straight line* so that we see how much the test mass is deviating from the *circle*

Such an ellipse has the form of a circle to a very good approximation as the solid ellipse in the left-hand picture of Fig. 9.12. However, the perfect ball is not quite in the center of the ellipse but rather somewhat shifted in the center of the *dashed* circle.

A test mass starts at the perihelion at point 1, heading parallel to the dashed circle, along the solid ellipse. Then it will *oscillate* about the dashed circle. At point 2, it already approaches the dashed circle. From point 3 on, it moves away from the dashed circle, and so on. After one year it passes through the starting point 1 at the same speed as it started, heading again parallel to the dashed circle.

We depicted this oscillation in the right-hand picture of Fig. 9.12 by expanding the dashed circle to a straight line.

We saw in Sect. 9.3 that Kepler's third law determines the period of a planet moving along the dashed circle. In other words: if the Newton law of gravity should be exactly correct, then Kepler's third law (9.16) would determine also the period of this orbital path.

Now we estimate how much the bending space-time changes the period of the oscillation when we see it from an inertial state far away from the central mass, as we did in Sect. 9.3 for Kepler's third law. We assume that the mass of the perfect ball is not too large so that similar to Sect. 9.2 we can **calculate perturbatively**. For this purpose we flatten the bended space-time near the dashed circle.

The arrows in Fig. 9.12 point at the direction of the radius. We know from the Schwarzschild exact solution Sect. 6.5 that the length of the vertical arrows will be *larger* than what the measuring tape shows, and that by the factor γ^{-1} of the radius of the dashed circle. Time is running more slowly by the γ factor, relative to us resting far away from the perfect ball in an inertial state. The speed at which the arrow changes its length is "change of length *per time*", hence it is larger than in the Newtonian theory, and that by the factor $\gamma^{-1}/\gamma = \gamma^{-2}$.

We can get the same effect by enlarging *all* vertical distances near the dashed circle by this factor γ^{-2}, but measuring the time with our clock. Of course this

Fig. 9.13 A test mass moving along the *solid ellipse* oscillates around the *dashed circular orbit* more slowly than Kepler's third law predicts. Therefore it turns more than 360 degrees between the perihelion and the next perihelion

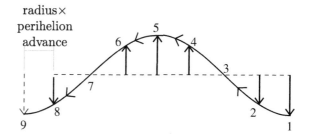

will lead to the incorrect space-time in general but near the dashed circle we have practically flattened out space-time. Because we measure time with our clock and gravity is not so strong, we can again use Kepler's third law (9.16) to get the period of the oscillation but with the radius enlarged by the factor γ^{-2}.

$$\frac{4\pi^2 \times \left(\text{radius} \times \gamma^{-2}\right)^3}{\left(\begin{array}{c}\text{gravity}\\\text{constant}\end{array}\right) \times \left(\begin{array}{c}\text{mass of}\\\text{perfect ball}\end{array}\right)} = \left(\text{period} \times \gamma^{-3}\right)^2 \qquad (9.17)$$

The square of the period will then grow by the factor $\left(\gamma^{-2}\right)^3 = \left(\gamma^{-3}\right)^2$, as you see on the right-hand side of Eq. (9.17). Therefore the period itself will be larger than one year by the factor γ^{-3}.

Hence the test mass again arrives at a perihelion, starting from one perihelion after about $360 \times \gamma^{-3}$ degrees, as you see in Fig. 9.13. We know from the Schwarzschild exact solution (9.4) that the square of the γ factor is smaller than one by

$$\gamma^2 = 1 - \frac{S}{\text{radius}}$$

We also know that for the solar system, this γ factor is nearly one. Hence we can use Appendix A.3 and see that the γ factor itself is about half of that amount smaller than one, that is by $1/2 \times S/\text{radius}$ and the inverse of its third power is about three times of that amount *larger* than one. Hence the **perihelion** of an ellipse will **advance** each year by the **angle**

$$\text{perihelion advance} = 360 \text{ degrees} \times \frac{3}{2} \times \frac{S}{\text{radius}} \qquad (9.18)$$

Let us check this for Earth. Earth moves nearly in a circle around the sun. So we can use the formula (9.18). Earth is at the distance 1.5×10^{11} from the sun. This is nearly the radius because gravity of the sun is weak. Hence the perihelion of the ellipse advances per year about

$$360 \times \frac{3}{2} \times \frac{2.96 \times 10^3}{1.5 \times 10^{11}} \approx 1.066 \times 10^{-5} \text{ degrees}$$

Per hundred years, this piles up one hundred times to 1.066×10^{-3} degrees. Multiplying with 60 gives this in **arc minutes**, and again with 60 gives the perihelion advance in **arc seconds**. Hence the ellipse of Earth should rotate by

$$3.8 \text{ arc seconds per century}$$

This is what astronomers observe![5]

9.5 Strong Gravity Near Black Holes

In Sects. 9.2 and 9.4, we have seen how the theory of general relativity acts if gravity is weak. In both cases the reason was that the Schwarzschild radius $S \approx 3000$ m is much smaller than the radius of the sun $\approx 6.96 \times 10^8$ m, and therefore much smaller than the smallest radius of the light beam or the planet moving around the sun. We can estimate the weakness of the gravity also from the speed of the test mass free-falling in a circle around the sun. According to the Kepler law (9.15), the squared speed in fractions of the speed of light is just half the fraction of the Schwarzschild radius and the radius of the circular path of the test mass,

$$\frac{\text{speed}^2}{c^2} = \frac{1}{2} \times \frac{S}{\text{radius}} \qquad (9.19)$$

For example, using the mean distance between Earth and sun as approximation for the radius 1.50×10^{11} m of Earth's orbit around the sun, the speed of the Earth around the sun is,

$$\frac{\text{speed}}{c} \approx \frac{1}{10000}$$

which is really much slower than the speed of light.

We see also from the Kepler law (9.19) that the nearer the test mass circles around the perfect ball, the faster it will be. If the perfect ball is a **black hole**, then how near can the test mass circle free-falling around the black hole?

The speed of the test mass entering the Kepler law is in terms of the time for us resting relative to the perfect ball in an inertial state. We know that no mass can pass

---

[5]See for example http://en.wikipedia.org/wiki/Tests_of_general_relativity.

Fig. 9.14 Testing how near
to a black hole that the black
test mass can free-fall around
it in a *circle*

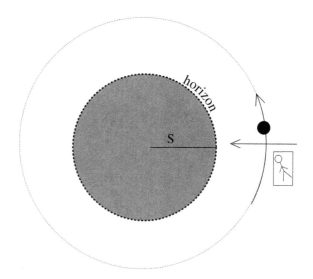

an observer at a speed faster than light. To let the test mass pass us in our inertial
state, we free-fall as usual from our place resting far away from the perfect ball, as
you see in Fig. 9.14.

The horizontal part of the speed at which the black test mass is passing us is the
speed of the Kepler law (9.19) at which the test mass is going around the black hole,

$$\frac{(\text{horizontal speed})^2}{c^2} = \frac{1}{2} \times \frac{S}{\text{radius}} \tag{9.20}$$

Its vertical part is our speed relative to the black hole, given by Eq. (9.3),

$$\frac{(\text{vertical speed})^2}{c^2} = \frac{S}{\text{radius}} \tag{9.21}$$

Our time runs at the same pace as the time in which we measured the Kepler law.
Hence we get the total speed at which the black test mass is passing us simply by
using the **Pythagoras theorem**, as in Fig. 9.15,

$$\frac{(\text{total speed})^2}{c^2} = \frac{1}{2} \times \frac{S}{\text{radius}} + 1 \times \frac{S}{\text{radius}} = \frac{3}{2} \times \frac{S}{\text{radius}} \tag{9.22}$$

The test mass can pass us at a speed that is not higher than the speed of light.
Hence the left-hand side of Eq. (9.22) is never larger than one, and so is the right-hand
side. In other words: any test mass free-falling on a circle around the black hole must
do so at a radius which is at least 1.5 times larger than the Schwarzschild radius. At
exactly 1.5 times the Schwarzschild radius, Eq. (9.22) tells us that a **light beam** will
go around the black hole in a **circle**!

Fig. 9.15 We get from the
Pythagoras theorem the total
speed relative to the
free-falling observer

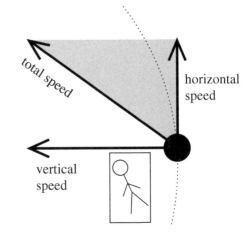

Fig. 9.16 A test mass
free-falling nearer to the
center of the black hole than
1.5 times the Schwarzschild
radius will pass the **horizon**,
that is it will free-fall into the
black hole

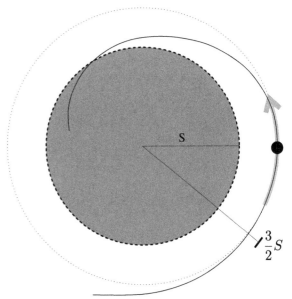

We can even say more: any free-falling test mass approaching the black hole nearer than 1.5 times the Schwarzschild radius will fall into the black hole, if it continues to free-fall. Why is that? If the test mass does not fall into the black hole, it will keep a smallest distance. Let us assume that the black test mass in Fig. 9.16 is just passing to the center of the black hole at its smallest distance. Hence it must move at *least* at the horizontal speed which it needs to continue on the dotted circular path. However, this dotted circle has a radius which is smaller than 1.5 times the Schwarzschild radius. That is impossible. Hence the black test mass can only escape if it leaves its free-falling state and starts for example a rocket motor to accelerate to an orbit which is at least 1.5 times the Schwarzschild radius away from the center of the black hole.

9.6 Gravitational Waves

No *body* can influence its neighborhood faster than light, not even with the help of gravity *itself*. We make a thought experiment to see what happens, if the bended space-time around two bodies is changing with time. For simplicity, let two *equal* masses swing back and forth under their gravity, and bouncing at each other *elastically* like billiard balls. We sketched their movement in Fig. 9.17. Observe that they both are *no* test masses. *Both* act with their gravity upon each other.

In the first line of the figure, the two balls start to move towards each other because of their gravity. We illustrated the gravity as a spring pulling the balls together.

In the second line, the balls move already at some speed. Now, gravity *itself* can only move at most at the speed of light. Hence, in order to reach the right-hand ball, the left-hand ball must have acted with its gravity already a little *earlier* so that the gravity can reach the right-hand ball at its position *right now*, and the other way round. We sketched the balls exercising their gravity in the past as dotted light gray balls and we let the spring begin and end there. However, the balls are then further apart and gravity acts *more weakly* than at their actual position.

In the third line, the balls collide and in the fourth line they bounce apart elastically. That is, they depart at the *same* speed as they collided some moments ago. In other words, they do *not* lose any of their motion energy due to the bouncing.

In the last line, gravity acts again a little bit earlier than at their actual position because it has to travel to influence the other ball. However, now gravity acts when

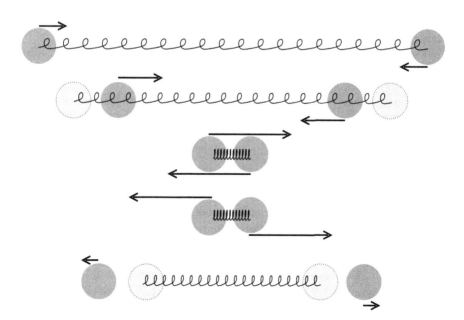

Fig. 9.17 Gravity acts weaker when the balls approach each other, than when they move away

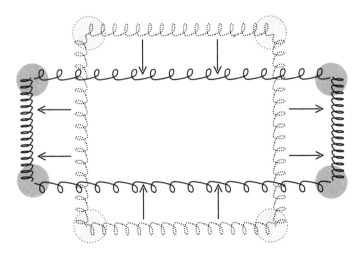

Fig. 9.18 As the gravitational wave is passing, the volume marked by the test masses is expanding in one direction and shrinking in the other direction

the balls were nearer than at their actual position, so it is *stronger* than at their actual position.

In a nutshell: the balls use more motion energy to overcome their gravity while departing from each other, than they gain from their gravity while approaching each other. In total, they have *less* energy than before when they are at rest again: they come nearer to each other at any time than when they began to approach each other because of **delayed gravity**.

Where has the energy gone? By bending space-time between the balls, gravity itself has not stopped when it reached the other ball, but spread. It formed **gravitational waves** because gravity acts at a distance *delayed*.

How can we measure a gravitational wave which is passing us? We saw in Fig. 5.12 that in vacuum, gravity does not contract the sufficiently small volumes but rather stretches them in one direction, and shrinks them in the other direction. Hence, when a gravitational wave is passing a volume marked with test masses, it should react as shown in Fig. 9.18. After the volume has deformed, some of the energy of the gravitational wave is in the springs between the test masses. This means that a gravitational wave is carrying energy.

Up to now, 2015, nobody has detected gravitational waves directly but Hulse and Taylor found in 1974 that two very compact stars, called PSR-1913+16, are moving around each other at a very close distance. Both have about the same mass of roughly 1.4 times the sun mass. They must have come from somewhere. Hence they lost energy by delayed gravity and therefore *they continue* to lose energy by spiraling ever faster towards each other. One of the two stars is a so-called **pulsar**, that is, it rotates very fast and emits electromagnetic waves such as the rotating mirror of a lighthouse acting as a timer. Physicists calculated with the help of the Einstein

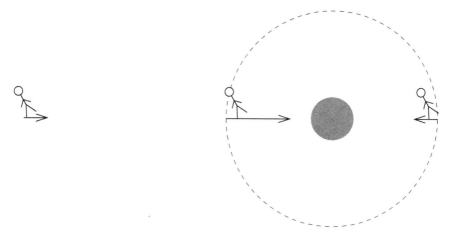

Fig. 9.19 Falling from different distances towards a perfect ball. The *arrows* show the direction and magnitude of the speed *relative* to the perfect ball

equation of gravity how much energy the two stars should lose to gravitational waves. Now, after 30 years of gathering data, we know that the theory of general relativity describes this effect correctly to at least 99 percent.[6]

9.7 Where Is the Gravity Energy?

Let us again free-fall vertically towards a perfect ball, as sketched in Fig. 9.19.

We start free-falling from the position at the far left. When we pass through the dashed imaginary circle, we are already traveling at some speed towards the perfect ball. At this instant, our friend on the right who had rested at this distance from the perfect ball begins to free-fall. The perfect ball moves with more speed towards *us* than it does towards our friend. Hence the perfect ball has *more* mass for us than for him, as we know from Sect. 2.7. However, according to the equivalence principle, we are in the same inertial state as we were when we started far away from the perfect ball. Hence we conclude that the perfect ball contains *more* mass when looked upon from a place resting far away from it than when looked upon from resting at some smaller distance. The energy of this mass seems to exist between us and the perfect ball. The question is only, where is this energy exactly?

Compare with the case of *elastic* energy. In Fig. 9.17 we used the picture that gravity attracts masses like a spring does. Maybe the space outside the perfect ball is bending such as an elastic body as the body sketched in Fig. 7.7 does? Inside the material we can imagine little springs holding the atoms together. When the material bends, the springs expand or contract, and thus contain the bending energy.

---

[6]See for example http://en.wikipedia.org/wiki/PSR_B1913+16.

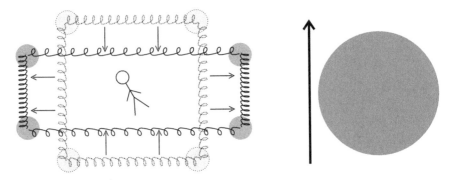

Fig. 9.20 As the perfect ball is passing, the volume marked by the test masses is expanding in one direction and shrinking in the other direction

Therefore let us check a small volume of space outside the perfect ball and search for the "gravity energy" in it. We mark the small volume by test masses which are resting relative to each other. However, when we release them, we know from the Einstein equation of gravity of Sect. 7.2 that the small volume will *not* begin to shrink because we are outside the mass of the perfect ball. Hence there cannot be any gravity energy inside this small volume which would have caused gravity.

However, in the case of gravitational waves in Sect. 9.6, we saw that test masses marking a small volume as in Fig. 9.18 *can* pick up energy from gravity in the form of a gravitational wave, even if it does *not* shrink the volume! We repeat this thought experiment now. We put test masses connected by springs in flat space-time and we mark a small volume being far away from other masses, as you see in Fig. 9.20. We are in the middle of the test masses. Then we let a perfect ball pass near us. While the perfect ball is passing, the test masses will react to the bending of space-time more or less as in the Fig. 5.12. When the springs have picked up some energy, we lock the springs until the perfect ball has passed.

The test masses and the springs then have more energy than before! What kind of energy did they pick up? The perfect ball moved *more slowly* away from us than it approached us. Hence they did *not* pick up any energy of bending space around the perfect ball but rather some small part of the *motion energy* of the perfect ball relative to us.

To sum up: precisely *because* bending space-time outside of a perfect ball does *not* shrink a small volume marked by test masses which are resting relative to each other, we cannot find the gravity energy in a small volume because we know that energy inside a small volume *would shrink* that volume. The gravity energy is in the *total* bended space-time around the gravitating mass but we *cannot* locate it! Why is that, *where* do you want to locate it? In *space* and *time*. That is the big difference: in Fig. 7.7, the *material* bends *in space-time* whereas gravity bends *space-time itself.*

9.8 Big Bang of the Universe

If we look at the night sky with the naked eye, we see planets, stars, and galaxies. Between them, there is the void. However, to the best of our knowledge the universe looks pretty much the same in all directions, if we only look at distances which are much larger than between galaxies. On such scales, there is everywhere the same amount of mass per volume. This observation is called the **cosmological principle**. What is more, only gravity seems to act over such large distances.

Hence we construct the following *simplest model* of the universe:

1. In the universe the energy including mass spreads everywhere and in every direction to the same extent.
2. Gravity dominates between the masses but no other forces and nearly no **pressure**.
3. The number of stars and galaxies in the universe does not change with time.

Then in what way can the energy in the universe bend space-time? In this model, time must run everywhere at the same pace on a coarse scale because mass spreads everywhere the same. Lengths can change but again lengths must change by the same amount *everywhere*, and on top of this, lengths must change by the same amount *in any direction*. The only way that this can happen is that lengths only *change with time*. This whole picture is called the **Friedman model** of the universe.

We illustrate this in Fig. 9.21. We sketched galaxies as stars. A typical length is the distance between two galaxies. *All* bodies move away from each other if time proceeds as we pass on from the left-hand to the right-hand picture or these bodies move towards each other if time proceeds as we pass from the right to the left.

The universe does not need to be infinitely large. Suppose that the universe had only two dimensions, such as the surface of a balloon. We visualize the stars and the galaxies by adhering small pips to it. If we blow up the balloon, all pips on the balloon will move away from each other, and there is no center, just as in Fig. 9.21. However, the pips *themselves* do not grow! In a similar way, if three dimensional

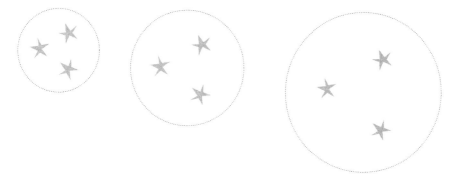

Fig. 9.21 The universe can expand or shrink, depending on whether time is proceeding as we pass from the left to the right, or vice versa

space around us grows, then from a certain large average distance, all galaxies will move away from us but the stars and galaxies themselves do not grow.

So we would like to find an answer to the following questions:

1. Can we not have a static universe or must it shrink or expand?
2. Whichever the case, can we *calculate* the expansion or shrinking rate?

We start by looking around us in our neighborhood and generalize to larger regions of the universe later on.

9.8.1 Small Ball of Mass in the Universe

We choose a small ball around us. In the real universe, "small" means "still large enough so that the mass and the energy spread more or less equally in all directions". We sketched the ball in Fig. 9.22.

The rest of the universe does *not* influence our small ball at all! Why is that? We assumed in the Friedman model that only gravity acts among the galaxies. What is more, mass spreads *in any direction* in the same way. Hence by the **Birkhoff theorem** of Sect. 8.3 the outside masses do not gravitate inside our small ball.

How does then the gravity of the masses act inside the ball? As in Sect. 8.2 we place black test masses on the surface of the ball which rest there relative to each other, and let them loose, so that they start to free-fall. Then the Einstein equation of gravity of Sect. 7.2 tells us:

Fig. 9.22 We placed black test masses at the rim of the ball, resting relative to each other. We just released them so that they free-fall as the nearby *stars* do, and therefore they do *not* accelerate relative to the *nearby stars*

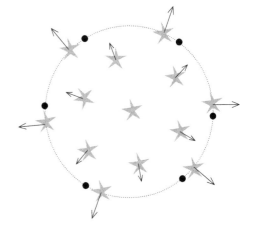

> The *relative* rate at which the *volume* of the small ball marked by the test masses begins to shrink is 4π times the gravity constant times the *density* of the mass in the ball.

Galaxies, stars and the like at the surface of the ball may move relative to the resting test masses but they do not **accelerate relative** to the nearby test masses because both free-fall. Hence we can let the ball move with the galaxies on its surface. This means that the rate at which the growth of the ball begins to slow down or the shrinking of the ball is quickening is the same as for the test masses. In other words: forget the test masses. The Einstein equation reads now

> The *relative* rate, at which the *volume* of the small ball—marked by the galaxies and moving with the galaxies on its surface—begins to slow down its growth or begins to accelerate its shrinking, is 4π times the gravity constant times the *density* of the mass of the ball.

We express this law now in terms of the *radius* of the small ball. In the Friedman model universe, we can make the ball as small as we wish because we assumed that mass spreads *everywhere* in the same way. This also means: there is everywhere the same amount of mass per volume and the smaller the ball is, the less mass is in it, and the less space-time is bending. Hence for a sufficiently small ball, it is nearly perfectly correct to use school geometry to calculate its volume in terms of its radius.

In school geometry, the volume of the small ball is in proportion to the *third* power of its radius. If for example the radius of the ball shrinks from 1 to 0.999, that is *relatively* by one per thousand, then its volume shrinks by $0.999^3 \approx 0.997$, that is *relatively* by three per thousand, three times as much as the *relative* shrinking of the radius.

Therefore the relative rate at which the growth of the volume of the small ball of galaxies begins to slow down, or the shrinking begins to accelerate, is *three* times the negative *acceleration* of the radius, per radius. The Einstein equation of gravity of Sect. 7.2 tells us that this is

$$-3 \times \frac{\text{acceleration of radius}}{\text{radius}} = 4\pi \times \left(\begin{array}{c} \text{gravity} \\ \text{constant} \end{array}\right) \times (\text{mass density}) \qquad (9.23)$$

This acceleration is, in the case of an expanding ball, the rate at which its expansion is slowing down.

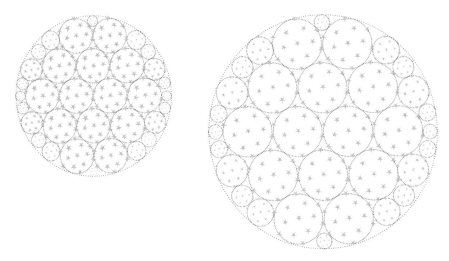

Fig. 9.23 The larger ball grows or shrinks *relatively* in the same manner as all the small balls grow or shrink

9.8.2 Large Ball of Mass in the Universe

We really wanted to know how much larger portions of the universe behave under gravity. In the Friedman model, mass spreads to the same extent everywhere so that even for a large ball of mass, as before, only the gravity of the masses *inside* this ball determines how much the ball will begin to shrink.

By the cosmological principle, the mass density on the right-hand side of the Eq. (9.23) is everywhere the same. Hence the radius of a ten times as large ball must begin to change *relatively* in the same way because we can fill it with smaller balls, all of which begin to change by the same *relative* amount as you see in Fig. 9.23. Therefore we can use this equation for a ball of *any size* in the universe!

We want to rewrite the gravity law for the universe in terms of the *mass* inside a large ball because in our model, the mass of the stars and galaxies inside a ball like in Fig. 9.22 does not change with time as the ball is growing or shrinking. However, the mass *density* will change, if the ball grows or shrinks. Mass density is mass per volume. So we can use that instead of the right-hand side of the Einstein equation (9.23). However, inside a large ball there is so much mass that space-time, and in particular space, will bend. However we cannot use school geometry. The volume of that large ball is *not* in proportion to the third power of its radius at a given time but there is a much more complicated law connecting the radius and the volume of a large ball.

However, there is a way out. Please again have a look at Fig. 9.23. We see that if the volume of the large ball grows by say 2 percent, the volumes of the small balls also grow or shrink *relatively* by that amount. Hence the *volume* of the large ball grows or shrinks *relatively* by the same amount as the volumes of the small balls.

We also learned at the end of the last section that the radius of a large and a small ball grows or shrinks *relatively* by the same amount.

Finally, we know that school geometry holds nearly perfectly for the small balls and that the volume of the small balls shrinks or grows *relatively* three times as much as their radius. Hence also the volume of the large ball will grow or shrink *relatively* three times as much as *its* radius,

$$\text{volume of ball} = \frac{4\pi}{3} \frac{(\text{radius})^3}{\text{constant}}$$

The constant is one for very small balls and we again get the formula from school geometry.

The constant does depend on the radius of the ball at *some* given time but does *not* change with time. Hence the mass density of the ball is in terms of the radius of the ball,

$$
\begin{aligned}
\text{mass density} &= \frac{\text{mass of ball}}{\text{volume of ball}} \\
&= \frac{\text{mass of ball}}{(\text{radius})^3} \times \frac{\text{constant}}{\dfrac{4\pi}{3}}
\end{aligned}
\qquad (9.24)
$$

We can use the Einstein equation of gravity (9.23) which we write here again,

$$-3 \times \frac{\text{acceleration of radius}}{\text{radius}} = 4\pi \times \left(\frac{\text{gravity}}{\text{constant}}\right) \times (\text{mass density})$$

and replace the "mass density" in terms of the mass and the radius of the large ball of Eq. (9.24),

$$-\cancel{3} \times \frac{\text{acceleration of radius}}{\cancel{\text{radius}}} = \frac{\cancel{4\pi} \left(\dfrac{\text{gravity}}{\text{constant}}\right) \times (\text{mass of ball})}{(\text{radius})^{\cancel{3}\,2}} \times \frac{\text{constant}}{\dfrac{\cancel{4\pi}}{\cancel{3}}}$$

On both sides we can cancel one radius and the factors 3 and 4π, and arrive at

$$\text{acceleration of radius} = -\frac{\left(\dfrac{\text{gravity}}{\text{constant}}\right) \times (\text{mass of ball})}{(\text{radius})^2} \times \text{constant}$$

We can replace the radius with the distance to the center of the ball because in the Friedman model both are in proportion. Then the constant changes its value but the equation remains in the same form

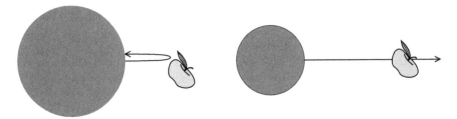

Fig. 9.24 The universe is expanding in the same way as an apple thrown vertically upwards from a small planet is slowing down

$$\begin{pmatrix} \text{acceleration of} \\ \text{distance to center} \end{pmatrix} = - \frac{\begin{pmatrix} \text{gravity} \\ \text{constant} \end{pmatrix} \times (\text{mass of ball})}{(\text{distance to center of the ball})^2} \times \text{constant} \quad (9.25)$$

This is the **Friedman equation**. We have seen this equation before: because time in the universe runs everywhere at the same pace, the equation looks just like the *Newton law of gravity* for a perfect ball (8.7), except for the numerical value of the constant. Let us coin a catch-phrase:

> For the universe as a whole, the law of gravity looks like the Newton law!

We can solve this equation in the opposite way than Newton did it: not by looking at an apple falling from a tree but by *throwing* an apple *vertically* into the sky. To get rid of the air drag, we do it as the little prince of Saint-Exupéry would do, that is on the surface of several small planets of different mass at the same initial speed. In Fig. 9.24 we sketched what happens: the planet on the left has *enough* mass so that the apple will eventually fall down. The planet on the right is so light that the apple is leaving it, moving away forever.

We have found again an **exact solution** of the **Einstein equation of gravity**!

Now we can answer the first question of p. 116: a *static* universe is *not stable*. It will immediately start to collapse, just as the apple resting above the planet will start to fall down. The universe has only two choices: to expand or to collapse. In fact, astronomers know since about 80 years that the distant galaxies all move away from each other. Hence they all must have been much more closer in the past. There must have been a **big bang** where all energy of the universe suddenly started to expand.

In the same way as for the apple on the small planet: if the universe has enough mass, then the expansion caused by the big bang will eventually stop and the universe will collapse in the future. If not, then the universe will forever expand, but in any case the expansion will *slow down*.

Therefore we should observe that the distant galaxies are moving *faster* away from us than the nearer galaxies because in the past the universe should have expanded

Fig. 9.25 Two light waves wiping each other out. In reality they are on top of each other, but we drew them beside each other for a better view

faster and slowed down its expansion up to today. However, in the last ten years or so astronomers found the *opposite* to be true. The universe *expands ever faster!*[7]

This is still a riddle but there is one possibility that we neglected: the *energy and the pressure of the vacuum.*

9.9 Vacuum Energy and Gravity

We have a definite idea about the **vacuum**. It is what is left over, after we removed everything. Hence nothing should be left, especially no energy. However, experiments show that energy *is* left over! Let us see how. Suppose two light waves travel through empty space, through the same path, but oscillating in the opposite way, as sketched in Fig. 9.25. Then in the same way as water waves, they wipe each other out. No wave is left over. However, both waves carry a *positive* amount of energy. Therefore no wave, but **vacuum energy** is left.

If such waves wipe each other out, how can we see them anyway? We can, and we now describe how they have been seen. Put two metal plates near to each other in vacuum. Metal does not allow light waves to pass, at least no waves with small enough wave lengths. Also, the light waves cannot oscillate inside the metal. In Fig. 9.26, we show in the left-hand picture oscillating light waves at one instant of time and in the right-hand picture some time later. Compare the two pictures: the upper two solid waves fit between the plates because near the plates they do not oscillate. However, the dashed wave does oscillate near the plates, so it cannot fit. This means that there are *more* possible waves and vacuum energy *without* metal plates than *with* metal plates.

Now, move the metal plates nearer together, say to one-fourth of their previous distance, as you see in Fig. 9.27. Then the upper wave does *not* fit any longer. Only the wave in the middle still fits. Hence *the more* the metal plates approach each other, *the less* energy fits between them. Therefore the metal plates begin to approach each other *by themselves* to lower their energy. This is what physicists observe! It is the **Casimir effect**.[8]

---

[7]This discovery got the Nobel prize in 2011. See for example http://en.wikipedia.org/wiki/Accelerating_universe.

[8]In 2001 this was observed by G. Bressi, G. Carugno, R. Onofrio, and G. Ruoso. You can download the original paper here: http://arxiv.org/abs/quant-ph/0203002.

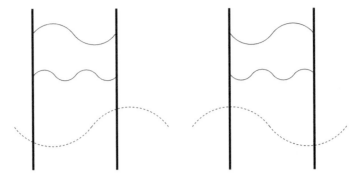

Fig. 9.26 Two parallel metal plates in vacuum, with oscillating light waves in between. We took the *right-hand* picture at some later time than the *left-hand* picture

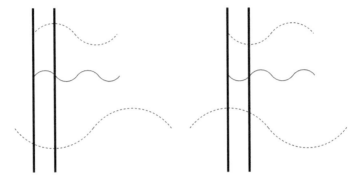

Fig. 9.27 The nearer the metal plates comes together, the fewer light waves fit in between

The Casimir effect is one of the most important phenomenons of the subatomic world, and we would need **quantum theory** to estimate it. This is beyond the scope of this book. In fact Casimir has done the calculation, and indeed quantum theory predicts the correct attraction between the two plates.

However, we can take the Casimir effect as an experimental fact showing that these "ghost" waves are real. By the way, they are called **vacuum fluctuations**. How much energy does such a light wave carry? Its energy is in proportion to its *inverse* wave length. The constant of proportion is the **Planck constant** times the speed of light.

We see that the more we look into small pieces of space, the more such waves with smaller and smaller wave lengths fit into that space. Because their energy goes as the inverse of their wavelength, this energy grows more and more! And *any* type of energy will gravitate!

Hence we have the same situation as we already discussed shortly in Sect. 7.3: the vacuum *itself* acts as a kind of gas in which matter immerses. We see from the Casimir effect that the **vacuum** is carrying energy, so it must exert **pressure**. Then

Fig. 9.28 Positive vacuum energy means negative pressure

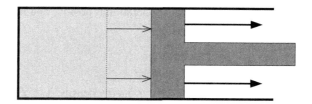

what pressure does the vacuum energy exert? For this, let us do a thought experiment that we *never* can do in reality. In Fig. 9.28, we sketched a lightly shaded volume containing vacuum with *nothing, not even a vacuum* around. Of course, that is why we never can realize this experiment, but never mind. We saw that vacuum has some definite positive energy. Hence to create more vacuum, we need to add energy. That means, we have to do some work to pull the piston to the right.

Compare to positive air pressure: we have to do work to *compress* the tire of a bicycle. So, for vacuum it is the other way round: it has **negative pressure**. In contrast to what we learned about the energy density in a pressure cooker in Sect. 1.12, we conclude that the pressure is here just the *negative* vacuum energy density itself. This pressure is the same in any of the three directions of space. The complete Einstein equation of gravity of Sect. 7.3 tells us that instead of the energy density of the vacuum we have to use

$$\left(\begin{array}{c}\text{energy density}\\\text{of vacuum}\end{array}\right) + 3\left(\begin{array}{c}\text{pressure} = \text{negative}\\\text{energy density}\\\text{of vacuum}\end{array}\right) = (-2)\left(\begin{array}{c}\text{energy density}\\\text{of vacuum}\end{array}\right)$$

This is a *constant, negative* mass density if we divide by c^2 using Eq. (1.12). Hence our equation for the universe expansion (9.23) gets a new source of acceleration,

$$-3\frac{\left(\begin{array}{c}\text{extra acceleration}\\\text{of radius}\end{array}\right)}{\text{radius}} = 4\pi\left(\begin{array}{c}\text{gravity}\\\text{constant}\end{array}\right) \times (-2)\frac{\left(\begin{array}{c}\text{energy density}\\\text{of vacuum}\end{array}\right)}{c^2}$$

Because the distance from the center is in proportion to the radius, we can use it as well, and use the constant to correct for larger balls, as before:

$$-3\frac{\left(\begin{array}{c}\text{extra acceleration}\\\text{of distance to center}\end{array}\right)}{\text{distance to center}}$$

$$= 4\pi\left(\begin{array}{c}\text{gravity}\\\text{constant}\end{array}\right) \times (-2)\frac{\left(\begin{array}{c}\text{energy density}\\\text{of vacuum}\end{array}\right)}{c^2} \times \text{constant}$$

In other words: this extra acceleration is positive and in proportion to the distance to
the center:

$$
\begin{pmatrix} \text{extra acceleration} \\ \text{of distance to center} \end{pmatrix}
$$

$$
= \frac{8\pi}{3c^2} \begin{pmatrix} \text{gravity} \\ \text{constant} \end{pmatrix} \times \begin{pmatrix} \text{energy density} \\ \text{of vacuum} \end{pmatrix} \times \begin{pmatrix} \text{distance} \\ \text{to center} \end{pmatrix} \times \text{constant}
$$

Hence the more the universe expands, the *larger* is this extra acceleration, while the
original slowing down of Eq. (9.25) is becoming *smaller*. Therefore already today
the vacuum energy should dominate the fate of the universe!

We say "should" because there is a snag. You can already guess that many, many
types of ghost waves fit in the universe. So there should be a huge amount of vacuum
energy, actually so large, that its huge negative pressure should explode the universe
right now! So there must be something else, some mechanism which keeps the vacuum energy much smaller. Nobody has any convincing explanation why the vacuum
energy is so small that we see the universe expanding so *slowly*. Or to put it another
way:

It seems that at present we do not understand how the effects of the subatomic
world fit into bended space-time, or how bending space-time influences the subatomic
world.

Chapter 10
Epilogue

We saw that the theory of relativity is based on four principles:

1. Light passes near an observer always at the same speed c.
2. Time, lengths, and all other steady speed have only a meaning relative to some observer.
3. The equivalence principle: a small enough mass moves under gravity, that is in bended space-time, free-falling, at a steady proper time.
4. Mass bends space-time in the simplest possible way which fits in with the first three principles. The rate, at which the volume of a small enough resting cloud of small pieces of matter begins to shrink, is in proportion to the *mass* in that cloud.

The first two principles are the foundation of the theory of special relativity, and all together form the theory of general relativity.

The theory does not ask what makes masses move technically or what atomic structure mass has but it explores how mass, momentum, energy on the one hand, and time and space on the other hand, interact. It even absorbs one force of nature, that is gravity into bended space-time so that gravity turns out to be no force at all. Einstein created this theory using only meticulous physical reasoning. All that makes the theory of relativity so beautiful.

We saw that especially the *equivalence principle* is a powerful tool to see how mass is *reacting* on bended space-time. It is not only the famous elevator allegory of Sect. 5.1, repeated in many popular books. Because mass acts *and* reacts on bended space-time, the equivalence principle helps to solve the Einstein equation of gravity or to derive the Friedman model of the universe as an example.

Checked again and again over a period of more than hundred years, the theory of relativity serves today as a *frame* in which more detailed theories of matter have to *fit*. For example, electrodynamics describing all electrical and magnetic processes fits naturally in from the beginning.

It took a long time to fit in **quantum theory**, that is the theory of the microcosm, with the theory of **special relativity**. Indeed, it would need at least one more book

© Springer International Publishing Switzerland 2015
K. Fischer, *Relativity for Everyone*, Undergraduate Lecture Notes
in Physics, DOI 10.1007/978-3-319-17891-2_10

to describe the weird, yet real phenomena of the quantum theory such as anti-matter, or why the **Higgs boson** found in 2013 is so important for our understanding of the relation between quantum theory and the theory of special relativity.

It remains for future generations to merge this theory of the microcosm with the theory of **general relativity** describing the largest structures in our world.

Appendix

A.1 Important Numbers

While the numbers in Table A.1 are in fact known to higher precision, they are here rounded to two decimal places.

A.2 Inertia of Pure Energy in Detail

Please have a look at Fig. A.1.

While building up, the **pure energy**, that is the light package **presses** against the wall. Pressure itself is force per area of the wall. So let us fix the area of the right side of the wall to be just one square-meter so that pressure and force are the same. During the time the light is pressing against the wall, the wall receives a "push", that is **momentum**. This recoil momentum is mass times the speed of the wall, as we saw in Sect. 2.7. Double the time or double the pressure, and you will get twice as much recoil from the wall. Hence the recoil is the pressure of the light, times the time needed to build up the package of light,

$$(\text{wall mass}) \times (\text{wall speed}) = \text{pressure} \times \text{time}$$

By the way, currently we can measure this light **pressure** in the laboratory.[1] We saw in Sect. 1.12 that this light pressure is energy per volume. This volume is the surface area of the wall which we set to one, times the width of the light package. Hence the pressure times the time it acts is the energy of the light, per width of the light package, times the time it acts on the wall

$$\text{pressure} \times \text{time} = \frac{\text{energy}}{\text{width}} \times \text{time}$$

---

[1] See for example: http://en.wikipedia.org/wiki/Radiation_pressure.

© Springer International Publishing Switzerland 2015
K. Fischer, *Relativity for Everyone*, Undergraduate Lecture Notes
in Physics, DOI 10.1007/978-3-319-17891-2

Table A.1 Important quantities and their units

Name of the number	Value	Unit
Speed of light c	$2.99792458 \times 10^8 \approx 3.00 \times 10^8$	m/s
Gravity constant G	6.67×10^{-11}	$m^3/(kg\,s^2)$
Sun's mass	1.99×10^{30}	kg
Sun's radius	6.96×10^8	m
Sun's Schwarzschild radius	2.96×10^3	m
Earth's mean distance from sun	1.50×10^{11}	m

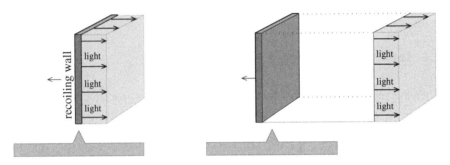

Fig. A.1 A wall recoils to the *left* because pure energy left the wall to the *right*

The package of light builds up with the speed of light, so that $\dfrac{\text{width}}{\text{time}}$ is just c,

$$\text{pressure} \times \text{time} = \frac{\text{energy}}{c}$$

Hence the wall will move to the left with speed

$$(\text{wall mass}) \times (\text{wall speed}) = \text{pressure} \times \text{time}$$

$$(\text{wall mass}) \times (\text{wall speed}) = \frac{\text{energy}}{c} = \frac{\text{energy}}{c^2} \times c \qquad (A.1)$$

After "some time" has elapsed, the wall traveled to the left to some distance which is the product (wall speed) × (some time). During this "some time", the light package traveled some distance $c \times$ (some time) to the right. Hence we multiply the Eq. (A.1) for the wall speed with this "some time", and we know how far the wall has traveled in the right-hand picture of Fig. A.1,

$$(\text{wall mass}) \times (\text{wall distance}) = \frac{\text{energy}}{c^2} \times (\text{light package distance})$$

The center of the total mass is still at rest. Hence while the wall carries its mass to the "wall distance" to the left, the light, that is the **pure energy** carries some **mass**

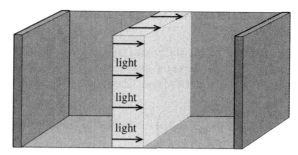

Fig. A.2 A packet of light moves from the *left* to the *right* wall of a *box*

to the "light package distance". Hence we see from the previous equation that this is the energy of the light package, divided by c^2,

$$\text{mass of light package} = \frac{\text{energy}}{c^2} \tag{A.2}$$

This is just Eq. (1.2).

We made several small mistakes. If the light carries some mass away from the wall, the wall must have lost this mass. However, we can make this error as small as we want by making the wall as massy as we wish. Likewise, the wall needed some time to react because the left part of the wall cannot know that light has already left the right part. However, we can think of the wall being as thin and dense as we wish.

In 1905 Einstein said[2]:

Wenn die Theorie den Tatsachen entspricht, so überträgt die Strahlung Trägheit zwischen den emittierenden und absorbierenden Körpern.

That is to say:

If the theory of relativity is correct, then radiation carries inertia between the emitting and absorbing body.

Apropos "absorbing body": in the original setup by Einstein the wall was the left wall of a box, and the right wall of the box absorbed the light again, as in Fig. A.2. He basically made the same calculation as we did here.

However, Einstein made the mistake to assume that this *whole* box recoiled *from the emitted light without delay*. This contradicts the theory of relativity because the right wall must then react *before* the light package reaches it, that is, it must have received the forces to react at a speed *faster* than light! You see that sometimes even Einstein himself had the wrong intuition about relativity.

---

[2] A. Einstein. Ist die Trägheit eines Körpers von seinem Energieinhalt abhängig? Annalen der Physik, Volume 18, page 639, 1905. A. Einstein. Das Prinzip von der Erhaltung der Schwerpunktsbewegung und die Trägheit der Energie. Annalen der Physik, Volume 20, page 627, 1906.

A.3 Relativity for Small Speeds

We know that relativistic effects nearly always depend on the speed via the γ factor. For small speeds, we know that the γ factor is nearly one. We want to know to what extent it is different from one. Check by hand or with a pocket calculator that

$$\left(1 - \frac{1}{2} \times 0.01\right)^2 = 0.995^2 = 0.990025 \approx 0.99 = 1 - 0.01$$

$$\left(1 - \frac{1}{2} \times 0.001\right)^2 = 0.9995^2 = 0.99900025 \approx 0.999 = 1 - 0.001$$

so that by extracting roots on both sides we have the estimate

$$1 - \frac{1}{2} \times 0.001 \approx \sqrt{1 - 0.001} = (1 - 0.001)^{\frac{1}{2}}$$

and we see that this estimate becomes the better, the less the γ factor differs from one. Therefore we can now estimate the γ factor at small speed

$$1 - \frac{1}{2}\left(\frac{\text{speed}}{c}\right)^2 \approx \sqrt{1 - \left(\frac{\text{speed}}{c}\right)^2} = \gamma \qquad (A.3)$$

The same pattern holds for other powers. If a number like 0.999 differs from one only by a very small amount -0.001, then the power of that number differs from one by that amount -0.001 times that power, if the power is not too large. For example, for the minus first or minus third power, we have

$$(1 - 0.001)^{-1} = \frac{1}{0.999} \approx 1.001001 \cdots \approx 1.001 = 1 + (-1) \times (-0.001)$$

$$(1 - 0.001)^{-3} = \frac{1}{0.999^3} \approx 1.0030006 \cdots \approx 1.003 = 1 + (-3) \times (-0.001)$$

and so on. Using the previous estimate (A.3), we have an estimate for the inverse γ factor at small speed,

$$\gamma^{-1} = \frac{1}{\gamma} \approx \frac{1}{1 - \frac{1}{2}\left(\dfrac{\text{speed}}{c}\right)^2} \approx 1 + \frac{1}{2}\left(\frac{\text{speed}}{c}\right)^2$$

$$\gamma^{-3} = \frac{1}{\gamma^3} \approx \frac{1}{1 - \frac{3}{2}\left(\dfrac{\text{speed}}{c}\right)^2} \approx 1 + \frac{3}{2}\left(\frac{\text{speed}}{c}\right)^2 \qquad (A.4)$$

A.4 Speed Addition from Growing Mass

From Fig. A.3 we see that the two balls of **resting-mass** m_0 have the total mass *relative to the box,*

$$M_0 = \frac{2m_0}{\sqrt{1 - \dfrac{u^2}{c^2}}}$$

because the square of both speeds $u^2 = (-u)^2$. This mass M_0 is the *resting-mass* of the box because we assumed the box itself to be very light so that the balls make up nearly all of its mass. Hence, when the box moves at the speed v, its mass should be

$$\frac{M_0}{\sqrt{1 - \dfrac{v^2}{c^2}}} = \frac{2m_0}{\sqrt{1 - \dfrac{u^2}{c^2}} \times \sqrt{1 - \dfrac{v^2}{c^2}}} \tag{A.5}$$

We check this mass by adding up the masses of the balls one at a time. If one ball moves in the box with a fraction $\dfrac{u}{c}$ of the speed of light and the box moves relative to the ground with a fraction $\dfrac{v}{c}$ of the speed of light, then we claim that the ball moves relative to the ground not with a fraction $\dfrac{u}{c} + \dfrac{v}{c}$ of the speed of light but at a smaller **total speed** w, whose fraction $\dfrac{w}{c}$ of the speed of light is,

$$\frac{w}{c} = \frac{\dfrac{u}{c} + \dfrac{v}{c}}{1 + \dfrac{u}{c} \times \dfrac{v}{c}} \tag{A.6}$$

To see this, we calculate the γ factor for w:

Fig. A.3 The left ball moves in the light box at the speed u to the right, and the right ball at the speed $-u$ to the left, *relative to the box.* The box itself moves at the speed v to the right, *relative to the ground*

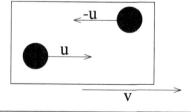

$$1 - \left(\frac{w}{c}\right)^2 = 1 - \left(\frac{\frac{u}{c}+\frac{v}{c}}{1+\frac{u}{c}\times\frac{v}{c}}\right)^2 = \frac{\left(1+\frac{u}{c}\times\frac{v}{c}\right)^2}{\left(1+\frac{u}{c}\times\frac{v}{c}\right)^2} - \frac{\left(\frac{u}{c}+\frac{v}{c}\right)^2}{\left(1+\frac{u}{c}\times\frac{v}{c}\right)^2}$$

We rearrange the numerator, crossing out the factors that cancel each other. We find that we can write it as the product of the γ factors of u and v:

$$\left(1+\frac{u}{c}\times\frac{v}{c}\right)^2 - \left(\frac{u}{c}+\frac{v}{c}\right)^2$$

$$= 1 + 2\frac{u}{c}\times\frac{v}{c} + \left(\frac{u}{c}\right)^2\left(\frac{v}{c}\right)^2 - \left(\frac{u}{c}\right)^2 - 2\frac{u}{c}\times\frac{v}{c} - \left(\frac{v}{c}\right)^2$$

$$= \left(1 - \frac{u^2}{c^2}\right)\times\left(1 - \frac{v^2}{c^2}\right)$$

Hence, extracting the root,

$$\sqrt{1 - \left(\frac{w}{c}\right)^2} = \frac{\sqrt{1 - \frac{u^2}{c^2}}\times\sqrt{1 - \frac{v^2}{c^2}}}{1+\frac{u}{c}\times\frac{v}{c}}$$

Therefore the ball traveling at the speed u inside the box traveling with the speed v attains the total mass

$$\frac{m_0}{\sqrt{1 - \frac{w^2}{c^2}}} = \frac{m_0\left(1+\frac{u}{c}\times\frac{v}{c}\right)}{\sqrt{1 - \frac{u^2}{c^2}}\times\sqrt{1 - \frac{v^2}{c^2}}}$$

Replacing u by $-u$ we arrive at the γ factor for the ball moving left at the speed $-u$ inside the box. This changes only the sign in the numerator from $1 + \frac{u}{c}\times\frac{v}{c}$ to $1 - \frac{u}{c}\times\frac{v}{c}$. Therefore the two moving masses add to

$$\frac{m_0\left(1+\frac{u}{c}\times\frac{v}{c}\right)}{\sqrt{1 - \frac{u^2}{c^2}}\times\sqrt{1 - \frac{v^2}{c^2}}} + \frac{m_0\left(1-\frac{u}{c}\times\frac{v}{c}\right)}{\sqrt{1 - \frac{u^2}{c^2}}\times\sqrt{1 - \frac{v^2}{c^2}}} = \frac{2m_0}{\sqrt{1 - \frac{u^2}{c^2}}\times\sqrt{1 - \frac{v^2}{c^2}}}$$

that is, to the same mass as in Eq. (A.5). Hence the Eq. (A.6) is the correct way to add parallel speeds.

A.5 Einstein Equation of Gravity in Terms of Tensors

The complete Einstein equation of gravity of Sect. 7.3 reads as follows:

The *relative* rate at which a small enough, resting cloud of matter begins to shrink grows in proportion to the *energy density* plus the *pressures* in each of the three directions in that cloud. The constant of proportion is 4π times the gravity constant G, divided by the square of the speed of light, or in short,

$$\begin{pmatrix} \text{relative shrinking} \\ \text{beginning rate} \end{pmatrix} = \frac{4\pi G}{c^2} \left[\begin{pmatrix} \text{energy} \\ \text{density} \end{pmatrix} + \begin{pmatrix} \text{sum of pressures} \\ \text{in each direction} \end{pmatrix} \right] \quad (A.7)$$

In order to be consistent with many textbooks, we write this equation in terms of tensors. Here we cannot give a course on tensor analysis but at least explain what the symbols mean. For matter resting relative to each other and us, energy density is one component of the **energy tensor** T_n^m. This tensor has sixteen components which are labeled by two indices m and n, both of which can take the values 0, 1, 2, or 3. Energy density is the component T_0^0. The *negative* of the component T_1^1 is the pressure in one direction of space, and $-T_2^2$ and $-T_3^3$ in the other two directions. For resting matter without internal stress, like a gas or a liquid, the other components are zero.

The beginning rate of the relative volume shrinking is the component R_0^0 of the **Ricci tensor**, times c^2. Hence the complete Einstein equation of gravity is

$$R_0^0 = \frac{4\pi G}{c^4} \left(T_0^0 - T_1^1 - T_2^2 - T_3^3 \right) \quad (A.8)$$

We still need a tensor U whose component U_0^0 gives the right-hand side of the equation. There is the tensor δ_n^m which is one if $m = n$ and zero else. Using the tensor

$$T = T_0^0 + T_1^1 + T_2^2 + T_3^3$$

which has only this one component T, we can build the tensor $T\delta_n^m$ and from it the tensor

$$U_n^m = 2T_n^m - T\delta_n^m$$

Its $^0_0$ component is

$$U_0^0 = 2T_0^0 - \left(T_0^0 + T_1^1 + T_2^2 + T_3^3 \right) = T_0^0 - T_1^1 - T_2^2 - T_3^3$$

That is precisely what we need for the right-hand side of Eq. (A.8). Hence the Einstein equation of gravity in terms of the energy tensor and Ricci tensor is

$$R_0^0 = \frac{4\pi G}{c^4} \left(2T_0^0 - T\delta_0^0 \right) \quad (A.9)$$

In Sects. 7.1 and 7.2, we chose the simplest case in which pressure is zero for a small cloud of test masses which are resting relative to each other. In this case, all components of the energy tensor are zero, except the component T_0^0 which gives the energy density. For more complicated distributed and moving masses, we need the equations for all the components of the Ricci and energy tensor. This is the **Einstein equation of gravity in terms of tensors,**

$$R_n^m = \frac{4\pi G}{c^4} \left(2T_n^m - T\delta_n^m\right) \qquad\qquad (A.10)$$

Index

A

Absolute speed of light, 4
Acceleration and equivalence principle, 48
Acceleration and free-fall, 46
Acceleration and gravity, 46
Acceleration and inertia, 6, 7, 37
Acceleration and light, 41, 66
Acceleration and school geometry, 40
Acceleration and straight ahead movement, 41
Acceleration by rotation, 37
Acceleration of radius, 88
Acceleration, relative, 61, 88, 92, 102, 117
Angle and perihelion advance, 107
$\approx$ means:is roughly equal to, ix
Arc minutes, 100, 108
Arc seconds, 100, 108

B

Bending angle of light beam, 96
Big bang, 120
Birkhoff theorem, 85, 116
Black hole, 68, 95, 108
Body speed and mass, 7

C

Casimir effect, 121
Causality, 25
CERN, 38
c for speed of light in vacuum, 4
Clock paradox, 38, 43, 61
Conserved quantity, 29
Cosmological principle, 115
Curvature of bended surface, 77

D

Delayed gravity, 112

E

Einstein about inertia and energy, 11
Einstein equation of gravity, 65, 73
Einstein equation of gravity, complete, 74
Einstein equation of gravity, exact solution of, 85, 90, 120
Einstein equation of gravity for energy, 73
Einstein equation of gravity in terms of mass density, 73
Einstein equation of gravity in terms of tensors, 134
Einstein law of gravity, 73
Einstein ring, 67
E is the symbol for energy, 11
Electrical charge is absolute, 30
Electric current makes magnets move, 30
Electric field, 32
Electric generator, 30, 31
Electric motor, 30, 31
Electrodynamics, 29, 32
Electromagnetic wave, 32
Ellipse, 101
$E = mc^2$, 12
Energy and information, 13
Energy conservation, 8
Energy-mass equivalence, 12
Energy tensor, 133
Energy, concept of, 8
Energy, kinetic = motion energy, 7
Energy, motion, 7
Energy, pure, 9
Energy, pure, mass of, 11, 128
Equivalence principle, 48

© Springer International Publishing Switzerland 2015
K. Fischer, *Relativity for Everyone*, Undergraduate Lecture Notes
in Physics, DOI 10.1007/978-3-319-17891-2

Euclidean geometry, 40
Exact solution of the Einstein equation of
    gravity, 85, 90, 120

F
Faraday paradox, 35
Faster than light, 5, 16
First law of relativity, 2, 34
Free-fall and reaction to gravity, 70
Free-falling, 46
Free-falling and bended space-time, 48
Free-falling and equivalence principle, 69
Free-falling and inertial state, 48
Free-falling and proper time, 52
Free-falling and straightest path, 52
Friedman-equation, 120
Friedman model, 115

G
Galilei, 2
Gamma factor, 16, 18
γ factor, 16, 18
Gamma factor, properties, 19
γ factor, properties, 19
Gauss curvature, 77
General theory of relativity: theory of gen-
    eral relativity, 80
Geodesic in space-time, 52, 70
Geodesic on surface, 50
Gravitational mass, 46
Gravitational waves, 112
Gravity bends space-time, 51
Gravity constant, 73
Gravity, delayed, 112
Gravity lens, 67
Gravity slows down clocks, 58

H
Higgs boson, 126
Horizon, 68, 95, 110

I
Inertia, 6
Inertia and time, 25
Inertial and gravitational mass are the same,
    46
Inertial mass, 6
Inertial mass under gravity, 65
Inertial state, 37
Inertial state and free-falling, 48

Inertial state and proper time, 42
Information and energy, 13

J
Joule, 8

K
Kepler laws, 101
Kepler law, third, for motion in a circle, 101
Kinetic energy = motion energy, 7

L
Law of motion, 81
Law of motion, Newton, 92
Left-hand rule, 30, 33
Length at right angles to movement, 22
Length in direction of speed, 21
Light as pure energy, 10
Light beam, going around a circle, 109
Light pressure, 127
Light, faster than, 5, 16
Lorentz force, 30, 32, 82
Lorentz force from electrical current, 33

M
Mach principle, 65
Magnetic field, 32
Mass, 6
Mass density, 73
Mass-energy equivalence, 12
Mass, gravitational, 46
Mass, inertial, 6
Mass, inertial under gravity, 65
Mass, resting, 9, 27, 131
Mass, total energy of, 27
Matter drifting in empty space and inertia,
    45
Maxwell equations, 32
Metric, 65
m is the symbol for mass, 11
Model of planet or star, 62
Momentum, 25, 127
Motion energy, 7
Muons, 38

N
Nearby path, in space, 51
Nearby path, in space-time, 52, 61
Negative pressure, 123

Newton law of gravity, 79, 91, 92
Newton law of motion, 92

O

On the Electrodynamics of Moving Bodies, 32

P

Path, nearby, in space, 51
Path, nearby, in space-time, 52, 61
Perfect ball, 62
Perfect ball and bended surface, 77
Perfect ball and exact solution of Einstein equation of gravity, 78
Perfect ball and Schwarzschild exact solution, 91
Perfect ball, gravity of, 64
Perfect ball, gravity outside of, 86
Perfect ball, hollow, 86
Perihelion, 101
Perihelion advance, 105, 107
Perturbative calculation, 97, 106
Planck constant, 122
Pressure and disordered motion of particles, 74
Pressure and gravity, 74
Pressure and model of universe, 115
Pressure and pure energy, 10, 127
Pressure of vacuum, 122
Proper time, 42
Proper time and bending space-time, 52
Proper time and inertia, 44
Pulsar, 112
Pure energy, 9
Pure energy and gravity, 73
Pure energy, mass of, 11
Pythagoras theorem, 17, 40, 109

Q

Quantum theory and black holes, 95
Quantum theory and general relativity, 125, 126
Quantum theory and vacuum, 122

R

Relativity, first law, 2, 34
Relativity, second law, 4, 35
Resting-mass, 9, 27, 131
Resting-mass and motion energy, 9
Resting-mass and pure energy, 9
Resting-mass and total mass, 26
Ricci tensor, 78, 133

Riemann curvature tensor, 77
Roughly equal to:≈, ix

S

School geometry, 39, 40
Schwarzschild exact solution, 65, 79, 90
Schwarzschild, Karl, 65
Schwarzschild metric, 65
Schwarzschild radius, 93
Second law of relativity, 4, 35
Shrinking rate of volume, relative, 73
Shrinking volume, 71
Space, small volume, shrinking in time, 72
Space-time, 42
Space-time, bended, 51
Space-time inside hollow perfect ball, 85
Space-time nearby path, 52, 61
Special theory of relativity: theory of special relativity, 37
Speed addition, relativistic, 28
Speed addition, relativistic: total speed, 131
Speed of light, absolute, versus time, 16
Speed of light, direction of, 15
Speed of light in vacuum, 1
Speed of light is absolute, 4
S stands for Schwarzschild radius, 93
Straightest path, 49
Straightest path and bending space-time, 52

T

Tensor analysis, 78
Test mass, 52
Theory of general relativity, vii, 80
Theory of relativity, starting point of, 4
Third Kepler law, 104
Thought experiment, vii
Time, coincidence and speed, 24
Time machine, 25
Total mass and resting-mass, 26
Twin paradox, 38, 43, 61

U

Units, ix

V

Vacuum, 121
Vacuum energy, 121
Vacuum fluctuations, 122
Vacuum, light traveling through, 4
Vacuum pressure, 122
Volume, shrinking, 71

Printed in the United States
By Bookmasters